Modern Perspectives in Theoretical Physics

K. S. Sreelatha · Varghese Jacob
Editors

Modern Perspectives in Theoretical Physics

80th Birthday Festschrift in Honor of K. Babu Joseph

Editors
K. S. Sreelatha
Department of Physics
Government College Kottayam
Kottayam, Kerala, India

Varghese Jacob
Department of Mathematics
Government College Kottayam
Kottayam, Kerala, India

ISBN 978-981-15-9312-3 ISBN 978-981-15-9313-0 (eBook)
https://doi.org/10.1007/978-981-15-9313-0

This Springer imprint is published by the registered company Springer Nature Singapore Pte Ltd.
The registered company address is: 152 Beach Road, #21-01/04 Gateway East, Singapore 189721, Singapore

Foreword by P. P. Divakaran

I first met Babu Joseph sometime in the mid-1970s in connection with the Ph.D. qualifying examination of one of his first students at the then not-so-old Cochin University of Science and Technology. After that, my occasional visits, for various purposes, continued over many years; especially memorable was the decade that followed immediately and I am now happy to remember my friendship and association with several of Babu Joseph's students, some of whom subsequently became stalwart members of the CUSAT physics faculty. It was an exciting time: several of the areas of research in which CUSAT became an active and respected participant were already beginning to produce new results. It was an exciting time also for the values of good science in Kerala. The hard struggle to save Silent Valley from those who would destroy our natural heritage was at its height, a battle in which CUSAT played its due role. My visits gave me, personally, a ringside seat from which to experience the power that a committed community of citizen-scientists can wield.

Babu Joseph's own high standards, both in the choice of problems on which he and his students worked and in the manner in which they were solved and written up, played no small part in making CUSAT such a stimulating place to be at. To this must be added his personal qualities: generousity, open-mindedness and hospitability. This is not just my personal experience; I know many visitors to CUSAT who will heartily endorse it.

Babu Joseph's main scientific interest has always been physics at its most fundamental level which, at that time, was concerned mostly with the description and understanding of the physics of elementary particles. The high point of that period was the organisation of one of the regularly held national high energy physics symposia (in the winter of 1980–81, if memory serves) by him. It was among the very best of these symposia, with a wide participation of distinguished high energy physicists from India (and a few from outside). Among the many memories I have of it, two stand out: scientifically, the several unofficial or informal sessions, full of passion and noise, devoted to a particular theme or a particularly interesting new piece of work; socially, the warm efficiency with which everything was organised and conducted.

Subsequently, Babu Joseph's urge to go to the roots of interesting intellectual/rational issues extended to areas of fundamental physics other than particle physics, quantum field theory, etc. and even to complex systems which are not generally considered part of physics. Along with that came an increased sense of responsibility to his institution where he served first as the head of his department and then as the vice chancellor. Everyone knows that the position of the vice chancellor of a university, starved of essential funding—especially if it is not a central university—and subject to pressure from various institutions and authorities, is not a bed of roses. But CUSAT survived and thrived, as seen most clearly in its collaboration with the Inter-University Centre for Astronomy and Astrophysics (Pune) in the seeding of a culture of top-level research and teaching in various educational institutions across Kerala. It is a tribute to Babu Joseph's role in promoting this very fruitful collaboration in the field of astronomy/astrophysics/cosmology, of which, Kerala must remain forever conscious.

After his retirement from CUSAT, what I think of as Babu Joseph's true vocation has come fully to the forefront of his life. His search for an intellectual understanding of the philosophical–historical foundations of modern science and his love of writing—and of the literary life in its broadest meaning—have come together in his life as a writer, the author of dozens of articles and several books in Malayalam. They have had an impact on Kerala's intellectual life, keeping the flame alive at a time when many in our nation were engaged in a futile search for "science" in the mythical productions of certain periods of our past. These endeavours culminated in his role in the founding and continued health of the Malayalam magazine Ezhuthu. The life of the mind is not dead.

To become 80 years old is not a handicap. Michelangelo was in his mid-80s when he designed and supervised the building of the great dome of St. Peter's Basilica in Rome. Nilakantha, our own great polymath from somewhat earlier, was 80 or more when he completed his masterpiece, his bhasya of Aryabhatiya. Onward!

P. P. Divakaran
Formerly Professor of Physics
TIFR, Mumbai

Foreword by M. Lakshmanan

Young people often look for a role model to shape their careers. It is more so for people in the field of science. Professor K. Babu Joseph has been one such towering personality for many youngsters in the southern part of the country, particularly for those who are pursuing scientific research. Professor Babu Joseph rose from the ranks of active young theoretical physicists in the early 1970s to a Professor of Physics and Head of the Department of Physics and then as a Vice Chancellor of Cochin University of Science and Technology (CUSAT), Kochi, to reach the pinnacle of his career. It is very apt that his grateful former research students have joined together to bring out this very useful festschrift 'Modern Perspectives in Theoretical Physics' on the occasion of his 80th birthday, and they should all be congratulated for their efforts.

Professor Babu Joseph's scientific interests have been vast and wide in theoretical physics. He explored deeply many areas such as quantum field theory, cosmology, nonlinear dynamics including solitons and chaos, mathematical physics and economics besides other topics and has published more than 150 research articles in his long academic career spanning more than four decades. He has trained more than 20 scholars towards their Ph.D. degrees. Many of them have emerged as leading theoretical physicists on their own and have led or are leading strong teams of active young researchers in different institutions of repute in the country. Professor Babu Joseph should be a satisfied person on this count.

Professor Babu Joseph has been a very successful administrator, both as Head of the Department as well as the Vice Chancellor of CUSAT during1997–2002. His calmness, uprightness, charming smile and friendly disposition towards all have endeared him as a great personality, and he is loved by one and all.

I had the good fortune of knowing personally Prof. Babu Joseph quite well since the late 1970s, and we have mutually visited each other at our institutes (and of course our homes as well). He has been an excellent host, and our research scholars had the benefit of visiting each others' institutions for scientific collaborations. I highly value his friendship and I am grateful to him for his kindness.

Professor Babu Joseph continues to be in active academic pursuit for the cause of higher education, particularly in the state of Kerala, even after his retirement as Vice Chancellor of CUSAT. I wish him continued good health, success and active service to the cause of science and higher education on his 80th birthday. Aspiring young students and researchers will always find Prof. Babu Joseph to be a model to be emulated.

M. Lakshmanan
Professor of Eminence & SERB
Distinguished Fellow
Department of Nonlinear Dynamics
Bharathidasan University
Tiruchirappalli, India

Preface

Professor K. Babu Joseph, who held the positions of teacher of physics, head of the department and later the Vice Chancellor of the prestigious Cochin University of Science and Technology (CUSAT), has promenaded into his octogenarian year. On his 80th birthday on 6th July 2020, we are greatly honoured to dedicate this Festschrift to Prof. K. Babu Joseph, our guru and mentor, to whom we all are immensely indebted for making our lives meaningful. This Festschrift is designed as a tribute to his distinguished achievements in various fields of physics and applied mathematics. This book is a compilation of research review articles by his research students to reflect his innovative thoughts that have made his professional life so rewarding. His contributions to popularizing science in the form of writings and lectures are greatly appreciated by the state and central governments.

Professor Babu Joseph is a versatile genius in Physics. His doctoral work was in the area of spectroscopy under the guidance of Prof. K. Venketeswarlu, the founder head of the Department of Physics, CUSAT, and former Professor and head of the Department of Physics, Annamalai University. Then he extended his research to domains of Theoretical Physics, like gravitation and cosmology, quantum field theory and high energy Physics, the areas of nonlinear dynamics—solitons, machine learning, neural networks, chaos, fractals and their applications, limited to not only Physics but also Biology and Economics.

This book is divided into four sections based on the contributions of Prof. Babu Joseph and his students.

The first section is on cosmology in which Moncy V. John, Suresh P. K. and Sivakumar C. present their research work during and after their doctoral research. Suresh P. K. has completed his Ph.D. under the guidance of the late Prof. V. C. Kuriakose, who was a student of Prof. K. Babu Joseph.

The second section deals with contributions to high energy physics and gravitation by Satheesh K. P. and Titus K. Mathew—a student of the late Prof. M Sabir, who, in turn, was a student of Prof. K. Babu Joseph. In the third section, contributions to Mathematical Physics by Vinod G. and Leelamma K. K. are enlisted. The fourth section is on nonlinear dynamics and its applications where the research works by Lakshmi Parameswar, Ninan Sajeeth Philip, Rose P. Ignatius and K. S. Sreelatha are included.

Along with the contribution to various branches of Physics, Prof. K. Babu Joseph is internationally renowned for his articles and books based on the fundamentals of mathematics and the development of conceptual Physics and their applications in various fields like the relation of the mind and the brain, evolution of species, variation of GDP and its effects, fuzzy logic and neural networking.

Professor K. Babu Joseph is widely appreciated for his associations with various journals, professional bodies and scholarly societies in science and humanities and possesses a seemingly tireless capacity for support and mentoring. We devote this Festschrift for his tangible love and unconditional support and motivation in all our activities.

Our success is due to the combined efforts of many people. We deeply acknowledge Prof. P. P. Divakaran, Tata Institute of Fundamental Research, Mumbai, and Prof. M. Lakshmanan, Centre for Nonlinear Dynamics, Bharathidasan University, for writing 'Forewords' for this Festschrift. We sincerely acknowledge the efforts of the members of the editorial board in sparing their valuable time and energy to compile the book in the LaTeX format. The indefatigable patience and hard work by Dr. Anoop K. Mathew and Dr. Jinchu I., Department of Physics, Government College, Kottayam, to combine all the articles in the present book format and to make our effort in its outright form are deeply appreciated. Special thanks to Suresh P. K., Prof. of Physics, University of Hyderabad, for the eye-catching design of the cover pages. We sincerely acknowledge the co-operation of each contributor to this Festschrift, and also for his/her critical comments and suggestions for enriching the content, presentation and organization of the book in its final form.

Kottayam, India
July 2020

K. S. Sreelatha
Varghese Jacob

A Tribute to Prof. K. Babu Joseph on His 80th Birthday

The mediocre teacher tells.
The good teacher explains. The superior teacher demonstrates.
The great teacher inspires.

William Arthur Ward

In many ways, the art of a teacher is to awaken a learning experience in realizing our true vitality. We all have had our fair share of personal experiences with Prof. K. Babu Joseph, and this book is a humble dedication to the person who has inspired us with the magic of Quantum Mechanics, High Energy Physics, Nonlinear Dynamics and Solitons. He has moulded us, taught us to fight every challenge and encouraged us to adapt to changes with a positive outlook. We are grateful to him for leading by example and enabling us to look up to him as a great mentor. He has left irreplaceable footprints in the areas of Cosmology and Mathematical Physics, especially in understanding Cosmological models and underlying cosmic parameters. Machine Learning for Artificial Intelligence-driven data science platforms is the current trend to predict many nonlinear phenomena. Being a true visionary, Prof. Babu Joseph has opened new pathways for data scientists and Artificial Intelligence enthusiasts via producing the first ever Ph.D. on this topic from a university in the country. His book Padartham Muthal Daivakanam Vare (DC Books) won the Science Literature award in the Science book (In-Depth Science) category in 2019. Furthermore, he has worked alongside legends such as Prof. E. C. G. Sudarshan and Prof. N. Mukunda and also associated with IUCAA Pune. It is indeed fascinating to be exposed to multidisciplinary aspects of research and life from our teacher and we are humbled to have interacted and learnt from his vast abode of expertise. It is an honour to present him this festschrift titled 'Modern Perspectives in Theoretical Physics' based on our work with him that includes excerpts from our research theses, manuscripts and book chapters. We dedicate this collection to Prof. K. Babu Joseph—a great teacher, most importantly a great human being.

Students of Prof. K. Babu Joseph

Contents

Nonlinear Dynamics

Contributors

Rose P. Ignatius Department of Physics, St. Theresa's College, Ernakulam, Kerala, India;
Department of Physics, Al-Ameen College, Edathala, Aluva, Kerala, India

Moncy V. John School of Pure and Applied Physics, Mahatma Gandhi University, Kottayam, Kerala, India;
Department of Physics, St. Thomas College, Kozhencherry, Kerala, India

K. S. Sreelatha Department of Physics, Government College Kottayam, Kerala, India

K. K. Leelamma Department of Physics, Union Christian College, Aluva, Kerala, India

Titus K. Mathew Department of Physics, Cochin University of Science and Technology, Kochi, Kerala, India

Lakshmi Parameswar Department of Physics, D. B. Pampa College, Parumala, Pathanamthitta, Kerala, India

Ninan Sajeeth Philip Artificial Intelligence Research and Intelligent Systems, Thelliyoor, Kerala, India;
Department of Physics, St. Thomas College, Kozhencherry, Kerala, India

K. P. Satheesh Government Brennen College, Thalassery, Kerala, India;
Institute for Intensive Research in Basic Sciences, M.G University, Kottayam, Kerala, India

C. Sivakumar Department of Physics, Maharaja's College, Kochi, Kerala, India

P. K. Suresh School of Physics, University of Hyderabad, Hyderabad, Telangana, India

G. Vinod Department of Physics, Sree Sankara College, Kalady, Ernakulam, Kerala, India

P. B. Vinod Kumar Department of Mathematics, Rajagiri School of Engineering and Technology, Kochi, Kerala, India

Cosmology

A Non-empty Bouncing Milne Model for the Universe

Moncy V. John

Abstract In the late 1990s, a paradigm shift occurred in cosmology when observations of Type Ia supernovae indicated that our universe is undergoing an accelerated expansion at present and that this is possibly due to the presence of some otherwise unknown, mysterious 'dark energy', in addition to ordinary matter and the already speculated 'dark matter'. Just before that, a theoretical model of the universe published by the present author and Prof. K. Babu Joseph predicted a universe expanding linearly (Milne-type coasting evolution), which had a bounce from an earlier contracting epoch. It suggested the presence of dark matter and dark energy in the universe, though with densities different from that estimated by the supernova search teams. It also suggested the possibility that matter is continuously being created in the universe at the expense of dark energy. In addition to broadly agreeing with all major cosmological observations, the model has the advantage that it has no cosmological problems. The linear coasting model has several interesting follow-ups in the literature, such as the $R_h = ct$ cosmological model which are reviewed. In the second part of the present paper, I present a complex extension of the classic Milne model of the universe and show that this can lead to a bouncing non-empty, coasting cosmological model as in the case of the above coasting model. In the new model, the universe had a previous contracting phase and it emerged to the present coasting expansion phase after a bounce that occurred during a very small time of the order of Planck time. The present model has the advantage that the space-time attains the usual Lorentzian signature for the metric tensor after the Planck epoch, though at the minimum radius corresponding to the bounce, the signature is Euclidean.

Keywords Physical cosmology · Coasting models · Quantum cosmology

---

M. V. John (✉)
School of Pure and Applied Physics, Mahatma Gandhi University, Kottayam, Kerala, India

Department of Physics, St. Thomas College, Kozhencherry, Kerala, India
e-mail: moncyjohn@yahoo.co.in

K. S. Sreelatha and V. Jacob (eds.), *Modern Perspectives in Theoretical Physics*,
https://doi.org/10.1007/978-981-15-9313-0_1

1 Introduction

A cosmological solution to Einstein's general theory of relativity [1–4] was obtained by A. Friedmann in the early 1920s. This first suggested an expanding universe, though the solution was not widely noticed at that time. It is interesting to note that even Einstein himself believed in the prevailing paradigm of 'static universe' during that period and did not heed to the prediction of general relativity that the universe is either expanding or contracting. Hubble's observation in 1929 that the redshifts of galaxies and their apparent magnitudes obey certain relation was interpreted as evidence to an expanding universe. This model was based on only two basic principles in addition to general relativity: (1) the universe is homogeneous and isotropic at large scales and (2) it contains only ordinary matter. But in the 1930s itself, astronomical observations such as that of galaxy rotation curves suggested that the visible matter is not the entire component of the universe. It may contain some invisible matter, which is even now called 'dark matter'. Such a model of the universe that contains ordinary matter, possibly along with some dark matter, provided a satisfactory picture of the universe with an initial singularity and a very hot early phase. This model is now famous as the hot big bang model and had several successes like the prediction of cosmic microwave background radiation (CMBR), which was actually observed in 1964. The two scientists, A. A. Penzias and R. W. Wilson, who made this milestone observation, were awarded Nobel prize in 1978.

The Friedmann model of the universe reigned till the end of last century, though not without opposition. An unmistakable prediction of this model was that the cosmic expansion gradually slows down (decelerated expansion) with time. A paradigm shift occurred in cosmology with the 1998 observation that standard candles such as Type Ia supernovae show exceptional dimming at very high redshifts and this turned the graph upside down and led to the notion that the universe is at present undergoing an accelerated expansion. The reason for this was speculated to be the presence of large amount of some unobservable, repulsive dark energy and this culminated in the currently popular Lambda-CDM model. Adam Riess, Brian Schmidt and Saul Perlmutter, who led the teams that made this observation were awarded Nobel prize in 2011 for their discovery.

Now, after two more decades, when the dust settles, is there evidence laid bare for a universe that is neither accelerating nor decelerating? Was the universe expanding linearly throughout its history, with a constant expansion rate? A universe that expands with a constant expansion rate is said to be 'coasting'. Such a straightening of the graph may mark yet another milestone in the development of cosmology. This is now discussed in a large number of papers published in leading journals of astronomy. In the first part of this paper, we discuss various coasting cosmological models, starting from the first such model by the British astronomer A. Milne. The Milne model is still considered in the literature as an interesting, albeit unphysical model. We also review briefly the first realistic eternal cosmological model with

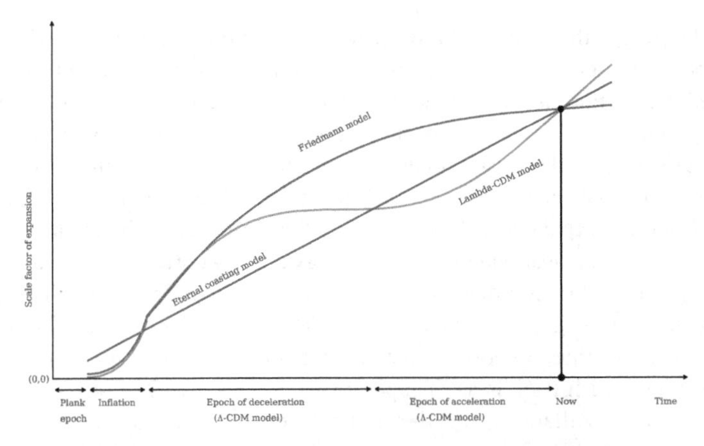

Fig. 1 Graphs showing expansion of the universe in different models (not to scale). Cosmic time is taken along the x-axis and the scale factor of expansion along the y-axis. There is no graph drawn for the Planck epoch. Following this, there occurs an exponential expansion for a short period in both Friedmann (green colour) and Lambda-CDM (red colour) models. This is referred to as the epoch of 'inflation'. Afterwards, the Friedmann model has only a decelerating phase, but the Lambda-CDM model has a decelerating phase followed by the recent accelerated expansion. The eternal coasting model (blue colour) has the most simple 'linear expansion' all throughout the history of the universe (On the time axis, various epochs for the Lambda-CDM model are marked.)

an initial bounce proposed and studied by the author, along with Professor K. Babu Joseph. In the second part, we present such a new realistic and physical eternal coasting cosmological model obtained through a complex extension of the Milne model (Fig. 1).

2 Eternal Coasting Models of the Universe

2.1 *Cosmological Problems*

Several problems, such as those referred to as the horizon problem, flatness problem, monopole problem, etc., were identified in Friedmann cosmology during the 1980s, but the theory of inflation [5] solved most of them at one stroke. But time and again, some other problems too came up in connection with this model. The theory of inflation, which speculates quantum field theoretical effects in the early epochs, suggests that the present universe is flat. In fact, the most important prediction of the inflationary models is that the universe is almost spatially flat. This in turn implies for a Friedmann model that the combination $H_0 t_0 \approx 2/3$, where H_0, the present Hubble constant is a measure of the expansion rate of the universe and t_0, the present age of the universe, i.e. the time elapsed since the big bang singularity. A major set back to inflationary models was in fact this prediction. Some observations in the mid-1990s [6] put this value to be lying in the range $0.85 < H_0 t_0 < 1.91$, contrary to the above prediction. Thus, when the condition of flatness is combined with the measured high value of the Hubble parameter, there arose an 'age problem'. The discovery of the accelerated expansion in 1998 predicted a cosmological constant Λ or some unknown dark energy along with ordinary matter, both these components having almost equal share in the present universe [7]. The importance of the discovery of accelerated expansion is that it resolved the age problem, but, in its turn, seeks an explanation to the near equality of the energy density corresponding to such Λ or dark energy and the matter density in the present universe. This is the 'coincidence problem' in cosmology, for which the ΛCDM model offers no satisfactory solution yet. Recently, three important cosmological observations, namely, the apparent magnitude and redshift of Type Ia supernovae (SN Ia), CMB power spectrum and baryon acoustic oscillations, together predict that $H_0 t_0$ of the present universe is very close to unity [8]. This is puzzling since according to the ΛCDM model, the product Ht could have values very different from unity; it can be anywhere in the range $0 < Ht < \infty$. Except during the period of inflation, in the past or future of the universe, this value shall not be unity either. Just like the coincidence problem, this synchronicity problem in the ΛCDM model is also ascribing certain special status to the epoch in which we live today.

2.2 Milne's Coasting Model

The Milne model [9] is a cosmological model widely discussed in the early stages of modern cosmology, proposed by the British physicist A. Milne. The model does not make use of the general theory of relativity. He does not include even gravity as an important interaction at the cosmic scale. Instead, Milne makes use of his theory of kinetic relativity, which is an extension of Einstein's special theory of relativity. Hence, its distinguishing feature is that all observers are in inertial motion at the cosmic level. However, it can also be viewed as a Friedmann model, but as having zero matter density, negative spatial curvature and the scale factor obeying the relation $a \propto t$. In this case, the expansion rate of the universe, which is measured as the Hubble parameter H, varies as $1/t$, the time elapsed since the big bang singularity. Hence, in this cosmological model, $H_0 t_0 = 1$ is not a problem; instead, this is its prediction. In [8], it is mentioned that the Milne model is the most suitable one to explain the above observational result, but the model is soon rejected on grounds that the model is empty and has negatively curved space sections (i.e. $\rho = 0$ and $k = -1$). This is quite reasonable, for an empty model is not a realistic one.

Very recently, an analysis of Type Ia supernova data [10] has appeared with the result that there is only marginal evidence for the widely accepted claim of the accelerated expansion of the universe. By a rigorous statistical analysis using the joint lightcurve analysis (JLA) catalogue of 740 SN Ia, it is found that the SN Ia Hubble diagram appears consistent with a uniform rate of expansion. This brings the Milne model again to the centre stage of cosmology, albeit in some new avatar.

A non-empty cosmological model which has a uniform expansion rate throughout the history of the universe may be called an 'eternal coasting model'. A characteristic feature for a coasting model is that it will have vanishing gravitational charge, i.e. $\rho + 3p = 0$, where ρ denotes the total matter/energy density of the universe and p is the pressure due to such matter/energy distribution. In view of the mounting supportive evidences obtained from the above analyses of most recent cosmological data [8, 10], one notes that the eternal coasting models deserve continued scrutiny, both from the conceptual and observational fronts.

2.3 *Non-empty Eternal Coasting Model*

A non-empty, closed cosmological model that is coasting throughout the history of the universe after a bounce from a previous contracting phase was first proposed in 1996 by the author and Professor Joseph [11, 12]. This model coincided with the widely discussed Ozer–Taha model [13, 14] at the earliest epochs. The attempt by Ozer and Taha was to put forward an alternative to the theory of inflation, which solves the flatness problem, horizon problem, etc. They obtained a bouncing, non-singular solution with $a = \sqrt{(a_0^2 + t^2)}$, and speculated that a_0, the minimum radius of the universe, has some small value. After the bounce, the model reaches the $a \propto t$ phase for some time, but soon deviates from it to enter the decelerating standard big bang evolution and continued in it. On the contrary, the eternal coasting model mentioned above in [11, 12] has the evolution $a = \sqrt{(a_0^2 + t^2)}$ throughout the cosmological history. Moreover, a quantum cosmological treatment gave the result that the minimum radius a_0 is of the order of Planck length. Also, the universe is comprised of visible matter, dark matter and dark energy and there is a possibility that matter is continuously being created in the universe at the expense of dark energy. These models were shown to have none of the cosmological problems, such as singularity, horizon, flatness, monopole, cosmological constant, size, age, etc. This has almost vanishing gravitational charge $\rho + 3p \approx 0$ after the bounce.

A slightly different model proposed by us in [15], which has $\rho + 3p = 0$ and an initial singularity, is an always coasting cosmology with $a \propto t$, obtained by extending the dimensional argument of [16]. This modified Chen–Wu model, which can have $k = 0, \pm 1$, has both matter and a time-varying cosmological constant. As in the above case, it consists of ordinary matter, dark matter and dark energy, which gradually decays to form matter. Comparison of this model with the ΛCDM model was performed using the SNe Ia data, with the help of the Bayesian theory [17], by the author and Professor J. V. Narlikar. The results showed that the evidence against the coasting model, when compared to the ΛCDM model, is only marginal. In 2005, an analysis was again performed [18] using the then-available SNe data to see whether the data really favours a decelerating past for the universe. Again, the conclusion was that the evidence is not strong enough to discriminate the ΛCDM model from a coasting cosmology. Recently, a quantum cosmological treatment [19] showed that a coasting solution is unique, since it has identical classical and quantum evolution, as in the case of free particles described by plane waves in ordinary quantum mechanics.

2.4 Other Coasting Models

In a well-known work, Kolb [20] has discussed at length the implications of a coasting cosmological model. In this, the universe is presently dominated by some exotic K-matter and is coasting, with equation of state $p_K = -\rho_K/3$. However, the model is different from the 'eternal coasting' models, for during the major part of its history in the early epochs, the universe is dominated by radiation and matter and hence its expansion rate is not uniform. Another cosmological model where the universe undergoes coasting evolution for part of its history was the one proposed by Sahni [21].

Some other realistic coasting models have also appeared in the literature since then. Nucleosynthesis in a different coasting model was discussed by Lohiya and co-workers in [22–24]. This issue is pursued in several works, including some recent ones [25]. They have also investigated the status of realistic coasting models with regard to other cosmological observations, such as the SNe Ia data [26].

Recently, another cosmological model named '$R_h = ct$ model' [27], which has an expansion history coinciding with that of the Milne model, is suggested as the true model of the universe [28, 29]. It is claimed that one-on-one comparative tests carried out between this model and the ΛCDM model using over 14 cosmological measurements and observations give conclusive evidence in support of the $R_h = ct$ model [30]. In the context of the latest synchronicity problem in cosmology and the analysis of JLA catalogue mentioned above, this model assumes great significance. Possibly due to this, in the literature, there are growing concerns regarding the fundamental basis of the theory itself [31, 32]. For instance, the latest paper on this subject [32] contends that $R_h = ct$ is a vacuum solution. Addressing this criticism, it is shown in [30] that the $R_h = ct$ model is not empty and hence not the same as the Milne model.

In a recent paper [33], it was pointed out that the $R_h = ct$ model is identical to the 'eternal coasting cosmological model' suggested by the author and Prof. Joseph [15]. Between these two models, there exist only some minor differences, and even these find their origin in certain vague and unsupported features assumed in the $R_h = ct$ model. The pre-existing models in [11, 12, 15] were proposed even before the release of the sensational SNe Ia data in 1998. The authors of the $R_h = ct$ model have performed almost the same kind of data analysis as in our case and reaches similar conclusions. Though these differently named models originated on different theoretical grounds, they all end up at the same cosmological solution for all practical purposes. It was specifically stated that the $R_h = ct$ model coincides with the coasting model in [15], for $k = 0$.

It was also pointed out that the only provision for the '$R_h = ct$ model' to claim any difference with the 'eternal coasting model' is with regard to variation of the densities of individual cosmic fluids. The authors of the former have stated that various constituents of the total energy can adjust their relative densities via particle–particle interactions [30] to get evolution equations that satisfy $p = -(1/3)\,\rho$. However, no specific time evolution for the densities of these constituents is suggested in the $R_h = ct$ papers. The authors excuse themselves by stating that when the evolution of individual components is needed, several conservation laws and reasonable assumptions delimit their behaviour [34]. Hence, there is no model here to compare with the eternal coasting model.

3 Complex Extension of the Milne Model

In this paper, we also show that the Milne model can lead to a bouncing non-empty, coasting cosmological model. This is done by a complex extension of scale factor in the Milne-type empty Friedmann model. The new model is realistic and nonsingular and is always coasting after the Planck epoch. As a next step, a quantum cosmological treatment of this model is made, which provides the result that the minimum radius a_0 in it is of the order of the Planck length. The complexified Milne model has Euclidean $(++++)$ signature at $t = 0$, but it changes to the usual Lorentzian one $(+\ -\ -\ -)$ immediately after the Planck epoch. We find that this leads to the widely speculated 'signature change' [35, 36] in the early universe.

Let us now extend the scale factor of the Milne model to the complex plane and denote it as $\hat{a}$. We can now show that this results in a coasting cosmology with real scale factor $a \equiv \mid \hat{a} \mid$. One can write the Friedmann equations corresponding to the empty Milne model (with units in which the speed of light $c = 1$) as

$$\left(\frac{\dot{\hat{a}}}{\hat{a}}\right)^2 - \frac{1}{\hat{a}^2} = 0 \tag{1}$$

and

$$2\frac{\ddot{\hat{a}}}{\hat{a}} + \left(\frac{\dot{\hat{a}}}{\hat{a}}\right)^2 - \frac{1}{\hat{a}^2} = 0. \tag{2}$$

Using the polar form $\hat{a} = ae^{i\phi}$ in these equations and then equating their real parts, we get

$$\frac{\dot{a}^2}{a^2} = \dot{\phi}^2 + \frac{1}{a^2}\cos 2\phi, \tag{3}$$

and

$$2\frac{\ddot{a}}{a} + \frac{\dot{a}^2}{a^2} = 3\dot{\phi}^2 + \frac{1}{a^2}\cos 2\phi. \tag{4}$$

These equations appear as the familiar Friedmann equations in inflationary models, for a universe with flat ($k = 0$) space sections and filled with a homogeneous scalar field ϕ. This model is now quite different from that of Milne, since the geometry of its space sections and the energy density are different. The right-hand sides of the equations imply the presence of a scalar field with kinetic energy $\dot{\phi}^2$ and potential energy $\cos(2\phi)/a^2$. That is, we have a new cosmological model with real scale factor $a =\mid \hat{a} \mid$, whose space sections are flat and which is non-empty. Equating also the imaginary parts, we get two supplementary equations

$$\ddot{\phi} + 2\dot{\phi}\frac{\dot{a}}{a} = 0 \tag{5}$$

and

$$2\dot{\phi}\frac{\dot{a}}{a} = -\frac{1}{a^2}\sin 2\phi.. \tag{6}$$

These shall be of help in solving the system of equations (3) and (4). However, one can directly solve Eqs. (1) and (2) to obtain $\hat{a} = (c_1 \pm ct) + ic_2$. Choosing the origin of time such that the constant of integration $c_1 = 0$ and relabelling $c_2 \equiv a_0$, we get the solution for $\hat{a}$ as

$$\hat{a} = \pm t + ia_0. \tag{7}$$

The solution for $a =\mid \hat{a} \mid$ can be seen to be

$$a = \sqrt{a_0^2 + t^2} \tag{8}$$

and the argument ϕ can be obtained as

$$\phi = \tan^{-1}\left(\frac{a_0}{t}\right). \tag{9}$$

This is a bouncing, nonsingular evolution with a_0 as the minimum value for the scale factor (Fig. 2). The time at which this minimum occurs for a is taken as $t = 0$. In the early phase of Ozer–Taha model [13, 14], the evolution is described by the same equation. If we take its energy density as comprised of matter and a time-variable dark energy and do not assume any arbitrary conservation equations for these individual components, all the cosmological problems mentioned above can be seen to vanish in it. The energy density ρ and pressure p in the new model, as can be deduced from Eqs. (3) and (4), have the following evolution. At the epoch near $t = 0$, the kinetic energy of the field ϕ is dominant, leading to the nonsingular behaviour. This phase of evolution is capable of solving cosmological problems such as that related to the horizon. For large t, we have $\rho \propto a^{-2}$ and $\rho + 3p \approx 0$. The contribution for ρ in the late universe comes almost entirely from the potential energy term. It may also be noted that its magnitude is very nearly equal to the critical density for the universe. One can explicitly write the variation of total density and pressure with scale factor as

$$\rho = \frac{3}{8\pi G}\left(\frac{1}{a^2} - \frac{a_0^2}{a^4}\right) \tag{10}$$

and

$$p = -\frac{1}{8\pi G}\left(\frac{1}{a^2} + \frac{a_0^2}{a^4}\right), \tag{11}$$

so that

$$\rho + 3p = -\frac{3}{4\pi G}\frac{a_0^2}{a^4}. \tag{12}$$

The term that causes the bounce to happen corresponds to a negative energy density, which can now be separated as

$$\rho_- = -\frac{3}{8\pi G}\frac{a_0^2}{a^4}. \tag{13}$$

The pressure due to this is given by $p_- = (1/3)\rho_-$, which is an equation of state characteristic of relativistic energy densities. This energy density becomes negligible for $a \gg a_0$, when compared to the rest of the energy densities (Fig. 2).

Thus, the real model is flat, with the total energy density and pressure obeying $\rho + 3p \approx 0$ for $a \gg a_0$. After this initial epoch, if the potential energy of the field comprises energy corresponding to radiation/matter and a time-variable dark energy, we can write $\rho = \rho_m + \rho_{d.e.}$. Taking $p_m = w\rho_m$ and $p_{d.e.} = -\rho_{d.e.}$, one obtains

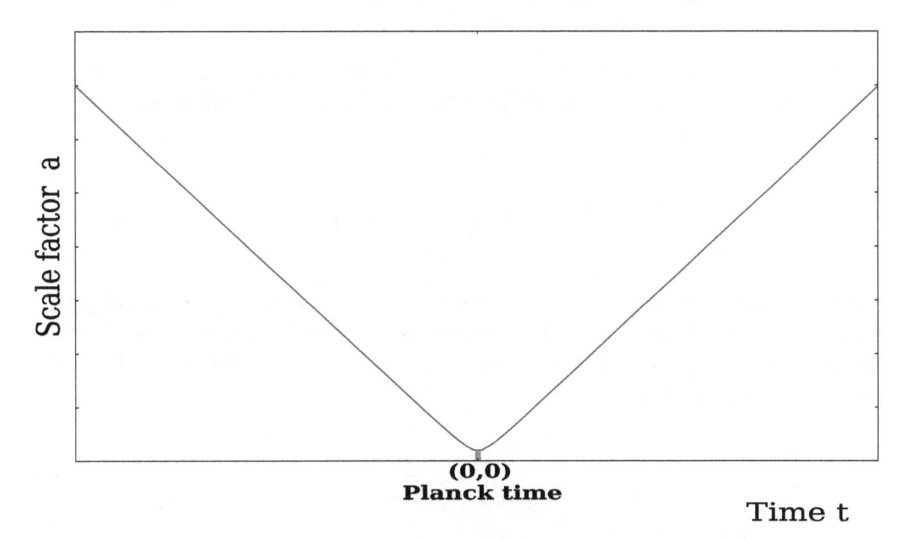

Fig. 2 The variation of scale factor with time in the non-empty bouncing Milne model is shown. The bounce occurs in between the coasting contraction and the coasting expansion phases and lasts only for 10^{-43} s, the Planck time

$$\Omega_m \equiv \frac{\rho_m}{\rho_c} = \frac{2}{3(1+w)}, \qquad \Omega_{d.e.} \equiv \frac{\rho_{d.e.}}{\rho_c} = \frac{1+3w}{3(1+w)}. \tag{14}$$

Since this universe is flat, the total density parameter $\Omega = \rho/\rho_c = 1$. When matter in the present universe is considered to be nonrelativistic with $w = 0$, the above equations predict $\Omega_m = 2/3$ and $\Omega_{d.e.} = 1/3$. The ratio between these densities is a constant of the order of unity throughout the expansion history (for $a \gg a_0$), and this avoids the coincidence problem. Being a coasting evolution, it naturally has no synchronicity problem. It solves all other cosmological problems, as demonstrated in the previous works.

With the aid of the solution (7), one can see that the complex-extended Milne model has Euclidean (++++) signature at $t = 0$, but it changes to the usual Lorentzian $(+ \; - \; - \; -)$ one for $a \gg a_0$. Hence, the complex extension of Milne model leads naturally to a signature change in the early universe, a possibility discussed extensively in the literature [35, 36]. The famous Hartle–Hawking 'no boundary' boundary condition [37] in quantum cosmology envisages a change of signature in the early universe. In this regime, there is no time and the space-time is purely spatial. Even in the classical Einstein equations, it is argued that the metric is Lorentzian not because it is demanded by the field equations; instead, it is a condition imposed on the metric before one looks for solutions [35]. In our case, the complex Milne model undergoes a signature change in a very natural way.

3.1 Wheeler–DeWitt Equation

In this section, we write down the Wheeler–DeWitt equation for the complex-extended Milne model. For this, we first note that the classical Milne model follows from the Lagrangian

$$L = -\frac{3\pi}{4G}\left(\frac{\dot{\hat{a}}^2\hat{a}}{N} + N\hat{a}\right). \tag{15}$$

Here N is called the lapse function, for which one can fix some convenient gauge. Writing down the Euler–Lagrange equations with respect to the variables N and $\hat{a}$ and fixing the gauge $N = 1$ leads to Eqs. (1) and (2), respectively. The canonically conjugate momentum is

$$\hat{\pi}_a = \frac{\partial L}{\partial \dot{\hat{a}}} = -\frac{3\pi}{2G}\frac{\dot{\hat{a}}\hat{a}}{N} \tag{16}$$

The canonical Hamiltonian can now be constructed as

$$\mathcal{H}_c = \hat{\pi}_a \dot{\hat{a}} - L = N\left(-\frac{G}{3\pi}\frac{\hat{\pi}_a^2}{\hat{a}} + \frac{3\pi}{4G}\hat{a}\right) \equiv N\mathcal{H}. \tag{17}$$

Quantization of a classical system like the one above means introduction of a wave function $\Psi(\hat{a})$ and requiring that it satisfies [38]

$$i\hbar\frac{\partial\Psi}{\partial t} = \mathcal{H}_c\Psi = N\mathcal{H}\Psi. \tag{18}$$

To ensure that time reparametrization invariance is not lost at the quantum level, the conventional practice is to ask that the wave function is annihilated by the operator version of $\mathcal{H}$, i.e.

$$\mathcal{H}\Psi = 0. \tag{19}$$

Equation (19) is called the Wheeler–DeWitt equation. It is analogous to a zero energy Schrodinger equation, in which the dynamical variable $\hat{a}$ and its conjugate momentum $\hat{\pi}_a$ are replaced by the corresponding operators. The wave function Ψ is defined on the minisuperspace with just one coordinate $\hat{a}$ and we expect it to provide information regarding the evolution of the universe. We may note here that the wave function is independent of time; it is a stationary solution in the minisuperspace.

The Wheeler–DeWitt equation for our case can be written by making the operator replacements for $\hat{\pi}_a$ and $\hat{a}$ in $\mathcal{H}$. However, finding the operator corresponding to $\hat{\pi}_a^2/\hat{a}$ is problematic due to an operator ordering ambiguity. We shall adopt the most commonly used form

$$\frac{\hat{\pi}_a^2}{\hat{a}} \to -\hat{a}^{-r-1}\frac{\partial}{\partial\hat{a}}\left(\hat{a}^r\frac{\partial}{\partial\hat{a}}\right), \tag{20}$$

where the choice of r is arbitrary and is usually made according to convenience. Using this expression with $r = -1$, we obtain the Wheeler–DeWitt equation for the complexified Milne model as

$$\frac{d^2\Psi}{d\hat{a}^2} - \frac{1}{\hat{a}}\frac{d\Psi}{d\hat{a}} + \frac{9\pi^2}{4G^2}\hat{a}^2\Psi = 0. \tag{21}$$

This equation has an exact solution

$$\Psi(a) \propto \exp\left(\pm i\frac{3\pi}{4G}\hat{a}^2\right). \tag{22}$$

Whether this solution corresponds to the classical evolution of the model can be determined by drawing the de Broglie–Bohm trajectories for this wave function [19]. Identifying $\Psi \equiv R\exp(iS)$, one can draw these quantum trajectories by using the de Broglie equation of motion

$$\hat{\pi}_a = \frac{\partial S}{\partial\hat{a}}. \tag{23}$$

In the present case of quantum cosmology, while using (16), this equation of motion simply reads

$$-\frac{3\pi}{2G}\dot{\hat{a}}\hat{a} = \pm\frac{3\pi}{2G}\hat{a} \quad \text{or} \quad \dot{\hat{a}} = \pm 1, \tag{24}$$

which is the same classical equation (1). Hence, here we have the same classical and quantum trajectories, implying identical behaviour as in the case of free particles described by plane waves in ordinary quantum mechanics.

A notable feature here is the appearance of the factor $\sqrt{2G/3\pi}$ in the wave function (22). This has value very nearly equal to the Planck length. In all quantum gravity theories, a natural length scale is the Planck length. Hence, one can deduce that the value of a_0, the imaginary constant appearing in the scale factor of the complex-extended Milne model is also of this value. In turn, this is the minimum radius of the bouncing, real, coasting model. Thus, we see that the nonsingular behaviour of the present coasting model is due to the quantum effects at the earliest moments within the Planck time.

3.2 Negative Energy Density

One can see that the complex extension of Milne model, which led to the non-empty, flat model with $a = \sqrt{a_0^2 + t^2}$, naturally brings in a negative energy which causes the bounce in the real model, and this is what relieves the model from the singularity problem. As can be expected, this negative energy has an equation of state corresponding to relativistic matter. There are speculations on a universe driven by a Casimir energy, which is negative [39]. Casimir first showed, on the basis of relativistic quantum field theory, that between two parallel perfect plane conductors separated by a distance l, there is a renormalized energy $E = \pi^2/720l^4$ per unit area [40]. For a static universe and for a massless scalar field, it was calculated that Casimir energy has a density [41].

$$\rho_{casimir} = -\frac{0.411505}{4\pi a^4}. \tag{25}$$

If we accept the value of $a_0 = \sqrt{2G/3\pi}$, which is nearly equal to the Planck length, our expression for negative energy (13) can be written as

$$\rho_- = -\frac{1}{4\pi a^4}. \tag{26}$$

Thus, there is some strong ground to believe that the negative energy appearing in the complex extension of Milne model is of the nature of Casimir energy. However, one must admit that a much deeper justification for this, on the basis of a true quantum field or similar theory, is needed here.

4 Discussion

The first coasting model for the universe, which was conceived by Milne, has been an attractive idea, primarily for its simplicity. Though it is an empty model and is devoid of any gravitational effects, the model is not forgotten even after three quarters of a century. The non-empty 'always coasting' cosmology in [11, 12] was proposed at a time when the universe was considered to be decelerating at the present epoch, i.e. well before the discovery of the unnatural dimming of SN Ia at large redshifts that led to the claim that the universe is accelerating. The solution of the cosmological problems, including the age and coincidence problems, was one of the strong motivations for its prediction. Presently, the model is more relevant in the context of the synchronicity problem, for the currently popular Λ-CDM model has no easy way out of it. The former model shall always remain a potential rival to the latter, unless more sophisticated data shows that $H_0 t_0 \neq 1$. (As mentioned by [8], if this value is very near to unity and is not exactly equal to it, there would be an even worse synchronicity problem.)

A caveat similar to that in the case of negative energy density is there also when the energy of the field ϕ is assumed to contribute to the energy of matter/radiation and the time-variable dark energy. Similar is the case for the creation of matter/radiation from dark energy, as envisaged in this model. One can only view these predictions as providing outlines of a future broader theory. But since we have the result from quantum cosmology that there is exact classical to quantum correspondence for this new model, the above pointers from the classical theory seem to be in the right direction.

References

1. P.J.E. Peebles, *Principles of Physical Cosmology* (Princeton University Press, Princeton, 1993)
2. S. Weinberg, *Gravitation and Cosmology* (Wiley, New York, 1972)
3. J.V. Narlikar, *Introduction to Cosmology* (1993)
4. J.A. Peacock, *Cosmological Physics* (Cambridge University Press, London, 1999)
5. A.H. Guth, Phys. Rev. D **23**, 347 (1981)
6. W.L. Freedman et al., Nature (London) **371**, 757 (1994)
7. M.J. Pierce et al., The Hubble constant and Virgo cluster distance from observations of Cepheid variables. Nature (London) **371**, 385 (1994)
8. A. Avelino, R.P. Kirshner, Astrophys. J. **828**, 35 (2016)
9. E.A. Milne (The Clarendon Press, Oxford, 1935)
10. J.T. Nielsen, A. Guffanti, S. Sarkar, Sci. Rep. (2016)
11. M.V. John, K.B. Joseph, Phys. Lett. B **387**, 466 (1996)
12. M.V. John, K.B. Joseph, Class. Quantum Gravity **14**, 1115 (1997)

13. M. Özer, M. Taha, Phys. Lett. B **171**, 363 (1986)
14. M. Özer, M. Taha, Nucl. Phys. B **287**, 776 (1987)
15. M.V. John, K.B. Joseph, Phys. Rev. D **61**, 087304 (2000)
16. W. Chen, Y.-S. Wu, Phys. Rev. D **41**, 695 (1990)
17. M.V. John, J.V. Narlikar, Phys. Rev. D **65**, 043506 (2002)
18. M.V. John, Astrophys. J. **630**, 667 (2005)
19. M.V. John, Gravit. Cosmol. **21**, 208 (2015)
20. E.W. Kolb, Astrophys. J. **344**, 543 (1989)
21. V. Sahni, H. Feldman, A. Stebbins, Astrophys. J. **385** (1995)
22. M. Sethi, A. Batra, D. Lohiya, Phys. Rev. D **60**, 108301 (1999)
23. D. Lohiya, M. Sethi, Class. Quantum Gravity **16**, 1545 (1999)
24. A. Batra, D. Lohiya, S. Mahajan, A. Mukherjee, Int. J. Mod. Phys. D **9**, 757 (2000)
25. P. Singh, D. Lohiya, J. Cosmol. Astropart. Phys. **061** (2015)
26. A. Dev, M. Sethi, D. Lohiya, Phys. Lett. B **504**, 207 (2001)
27. F. Melia, A. Shevchuk, Mon. Not. R. Astron. Soc. **419**, 2579 (2012)
28. F. Melia, Astron. J. **144**, 110 (2012)
29. F. Melia, R.S. Maier, Mon. Not. R. Astron. Soc. **432**, 2669 (2013)
30. F. Melia, Mon. Not. R. Astron. Soc. **446**, 1191 (2015)
31. M. Bilicki, M. Seikel, Mon. Not. R. Astron. Soc. **425**, 1664 (2012)
32. A. Mitra, MNRAS **442**, 382 (2014)
33. M.V. John, Mon. Not. R. Astron. Soc. **484**, L35 (2019)
34. F. Melia, M. Fatuzzo, Mon. Not. R. Astron. Soc. **456**, 3422 (2016)
35. G. Ellis, A. Sumeruk, D. Coule, C. Hellaby, Class. Quantum Gravity **9**, 1535 (1992)
36. S.A. Hayward, Class. Quantum Gravity **9**, 1851 (1992)
37. S.W. Hawking, Nucl. Phys. B **239**, 257 (1984)
38. E.W. Kolb, M.S. Turner, *The Early Universe* (Addison-Wesley, Redwood City, 1990)
39. R. Jaffe, Phys. Rev. D **72**, 021301 (2005)
40. H.B. Casimir, in *Proceedings of the KNAW*, vol. 793 (1948)
41. E. Elizalde, J. Math. Phys. **35**, 3308 (1994)

Cosmic Acceleration and Dark Energy

C. Sivakumar

Abstract A missing energy called dark energy in the universe is essential to account for many recent observations. An effective equation of state parameter that represents the cosmic fluid of the universe is postulated to predict the evolution of our flat universe which is presently expanding with an acceleration under the negative pressure of such a dark energy. Here a model is proposed to successfully describe the evolution of the universe after the inflationary epoch to a time when the size of the universe would be infinite, crossing the time of its shift from the phase of deceleration to acceleration using a simple equation of state consistent with the cosmic scenario.

Keywords Scale factor · Equation of state · Flat universe · Dark energy

1 Introduction

The development of GTR and cosmology is a major scientific success of the twentieth century. One of the solutions of Einstein's field equations is the Friedmann model, also called standard model in cosmology, which is based on cosmological principle. Recent observations suggest that ratio of energy density to critical density is one, that is, universe is flat. But the matter density including both baryonic and dark is only a small fraction of total density. So there is a missing energy called dark energy. Since 1998, when observations of supernovae Type-Ia pointed accelerated cosmic expansion, various observational measurements have established such evolutionary change in the history of the late universe. To find the source behind the late cosmic acceleration is certainly a major challenge for cosmologists.

The acceleration in the expansion of the universe is one of the remarkable discoveries in recent cosmological research. This acceleration requires that nearly three-quarters of the energy of the universe is in a component of fluid energy density with a negative pressure. Many dark energy models are proposed to explain this accel-

C. Sivakumar (✉)
Department of Physics, Maharaja's College, Kochi, Kerala, India
e-mail: thrisivc@yahoo.com

K. S. Sreelatha and V. Jacob (eds.), *Modern Perspectives in Theoretical Physics*,
https://doi.org/10.1007/978-981-15-9313-0_2

eration. The condition for late acceleration is that the equation of state parameter $\omega < -\frac{1}{3}$, where ω is the ratio of pressure to energy density of the fluid. In this article, we suggest that cosmic fluid is dark energy dominated at present called dark era and proposes an effective equation of state parameter for the cosmic fluid.

Einstein's field equations of general relativity: $G_{ik} = \frac{8\pi G}{c^4} T_{ik}$, when FRW metric is assumed, produce Friedmann equations [1]

$$\frac{\dot{R}^2}{R^2} - \frac{8\pi G\rho}{3} = \frac{-kc^2}{R^2}, \frac{\dot{R}}{R} = H \tag{1}$$

$$2\frac{\ddot{R}}{R} + \frac{\dot{R}^2}{R^2} + \frac{8\pi GP}{c^2} = \frac{-kc^2}{R^2} \tag{2}$$

Combining,

$$\frac{\ddot{R}}{R} = -\frac{4\pi G\rho}{3}(1 + 3\omega) \tag{3}$$

ρ is the mass density, P the pressure, R the scale factor, H the Hubble parameter, and k the curvature parameter of the universe. ρ includes density ρ_r of radiation, $\rho_{m(n)}$ of normal matter, $\rho_{m(d)}$ of dark matter, and ρ_d of dark energy, which, respectively, are 0, 5, 25, and 70% of the total, presently [2]. The brightest microwave background fluctuations that have been measured to the accuracy of 0.004 are about one degree across [2] to say that universe is flat. Using the apparent magnitude-redshift data for the distant Type-Ia supernovae, Hubble parameter has been fine-tuned to $2.173 \times 10^{-18}/s$ by Planck Mission [6]. The results of SCP establish the fact that $\ddot{R} > 0$ presently and it is nearly over 5 billion years since it started to accelerate [3, 4]. Also the negative pressure parameter for dark energy that makes the universe accelerate is at least 0.6 in size [5].

The equation of state needed to solve Friedmann equations is written as $P = \omega\rho c^2$, ω is the equation of state parameter. ω is taken as $\frac{1}{3}$ for radiation era or when the universe has only radiation, 0 for matter era or when only normal matter and it is -1 when the universe has dark energy like cosmological constant. Universe should expand with deceleration when $\omega = 0.333$ and 0 and it should expand with acceleration when $\omega < -0.333$ because of Eq. (3). Also, by energy conservation,

$$dU + PdV = 0 \tag{4}$$

where U is the energy of the universe and V the volume.

This leads to $\rho_r \propto R^{-4}$, $\rho_{m(n)} \propto R^{-3}$, if components are separately conserved.

2 The New Effective Equation of State Parameter

The fluid of universe has radiation, matter (both normal and dark), and dark energy in it and let the pressure, inspired by the changing pressure parameter and the decreasing acceleration of [8], be

$$P = \left(A - B\frac{R_0}{R}\right)\rho c^2 \tag{5}$$

where $P = \omega_{eff}\rho c^2$.

R_0 is the scale factor of the universe at the shift from the phase of deceleration to acceleration, ω_{eff} is the effective equation of state parameter of the total fluid pressure of the universe, and A and B are two dimensionless constants.

Then, integrating $d(R^3\rho) + \omega_{eff}\rho d(R^3) = 0$

$$H = \frac{1}{R^{1.5(1+A)}}\exp\left(1.5B\left(\frac{R_0}{R_i} - \frac{R_0}{R}\right)\right) \tag{6}$$

$R_i \sim 10^{26}$ is the scale factor of the universe at the end of cosmic inflation [2, 7] and H the Hubble parameter when the scale factor is R.

And Friedmann equations are

$$H^2 - \frac{8\pi G}{3}\left(\rho_r + \rho_{m(n)} + \rho_{m(d)} + \rho_d\right) = 0 \tag{7}$$

$$2\frac{\ddot{R}}{R} + H^2 + 8\pi G\left[A - B\frac{R_0}{R}\right]\left(\rho_r + \rho_{m(n)} + \rho_{m(d)} + \rho_d\right) = 0 \tag{8}$$

giving

$$\ddot{R} = -\frac{4\pi GR}{3}\left[1 + 3\left(A - B\frac{R_0}{R}\right)\right](\rho_r + \rho_{m(n)} + \rho_{m(d)} + \rho_d) \tag{9}$$

Friedmann equations with the equation of state and conservation of energy are quite enough to talk about the dynamics of the universe.

3 Results

Since the present Hubble number $H_p \sim 10^{-18}$, $R_p \sim \exp(B(\frac{R_0}{R_i} - \frac{R_0}{R}))(H^{\frac{12}{1+A}})$ (Table 1).

Let $A = -0.571$ expecting $R_p \sim 10^{26}$.

$\omega_{eff} = -0.333$ at $R = R_0$ in Eq. ((5)) has then $B = -0.238$.

$\omega_{eff} = \frac{1}{3}$ gives $\frac{R_0}{R_i} = 3.798$ that acceleration started when the size of the universe was 3.798 times the initial size.

Table 1 Choice of "A"

12/(1+A)	A
⋮	⋮
24	–0.500
26	–0.538
28	–0.571
30	–0.600
⋮	⋮
∞	–1.000

Table 2 Scale factor $R(\omega_{eff})$

ω_{eff}	R
+0.333	R_i
–0.333	R_0
–0.400	R_p
–0.571	∞

If the law of partial pressures is true for the cosmic fluid at least presently, with -0.6 for dark energy parameter,

$\omega_{effp} = 0.25\omega_{m(d)} - 0.42 \approx -0.4$ if the present dark matter parameter is that small and positive even if it contributes, gives $\frac{R_0}{R_p} = 0.718$ (-0.4 is a good fit as present parameter ω_{effp} is less than -0.333 but away enough from -0.571). The dark matter parameter could be thus around $+0.08$ (note that if $\omega_{dp} = -1.0$ as ΛCDM says, $\omega_{m(d)p} \approx +1.2$).

Substituting H_p in Eq. (6), $R_p = 5.420 \times 10^{26}$.

Note that order(R) is as per [2].

$\frac{R_0}{R_i}$ and $\frac{R_0}{R_p}$ give $R_i = 1.0 \times 10^{26}$ and $R_0 = 3.9 \times 10^{26}$.

Substituting H_p in Eq. (7),

$\rho_p = (\rho_{m(n)} + \rho_{m(d)} + \rho_d)_p = 8.455 \times 10^{-27} kg/m^3$

and $P_p = \omega_{effp}\rho_p c^2 = -3.046 \times 10^{-10} Pa$

Using A and B,

$\ddot{R}_p == +2.570 \times 10^{-10} s^{-2}$ (Table 2).

4 Conclusion

The model describes a universe that evolves during all the radiation era, matter era, and the present dark era. According to $\omega_{eff} = -0.571 + 0.238\frac{R_0}{R}$, $0.333 = \omega_{effi} \geq \omega_{eff} \geq -0.571 = \omega_{efff}$, $\omega_{efff} = -0.571$ is the restriction on ω_{eff} to protect the present accepted size $\sim 10^{26}$ of the universe. It is a flat one as required by the microwave background fluctuations measured by WMAP. Its expansion is consistent with the present value of Hubble number. It is presently accelerating as established by SCP, under the negative dark energy parameter and shift from the phase of deceleration to acceleration occurred in the past at the effective pressure parameter -0.333 as required by Friedmann equations. The size of the universe at the shift is about 0.718 times the present value (while the ΛCDM has a factor of 0.54). The acceleration of the universe, according to the model in the late time is decreasing—decreases to zero as the scale factor tends to infinite, unlike the acceleration of the standard model which increases linearly with scale factor saving the universe from spending an infinite amount of energy to produce infinite acceleration. The scale factor of the universe in the late time is approximately like $t^{\frac{3}{2}}$ while standard cosmology has an exponential function. And finally, it is important to see that it has the potential of talking about the size of the dark matter parameter.

A simple effective equation of state parameter for the cosmic fluid of the form described here can be a powerful alternative to the pressure parameter of the standard model in the light of its strength of predicting cosmological parameters in a simpler way. It seems the model is capable of describing the universe we are living in when the values are fine-tuned.

References

1. A. Friedmann, Gen. Rel. Gr. **31**, 1991 (1999)
2. WMAP. https://map.gsfc.nasa.gov/
3. S. Perlmutter, Ap. J. **517**, 565 (1999)
4. G. Adam, Riess The. Astron. J. **116**, 1009 (1998)
5. M. Peter, Garnavich ApJ. **509**, 74 (2011)
6. Planck 2015 results arxiv.org/abs/1502.01589 astro-ph (0674450) (2015)
7. A.H. Guth, Phys. Rev., D **23**, 347 (1980)
8. D.S. Hajdukovic, arxiv.0908.1047 To appear in Astrophysics and Space science (2016)

Signature of the Quantum Gravity on the CMB

P. K. Suresh

Abstract The quantum gravity effect on chaotic inflation reduces its tensor-to-scalar ratio than its predicted value and is found less than the estimated upper bound from the various CMB observations. The lowering nature of the power level of the lensed BB mode angular power spectrum, especially at lower multipoles, compared to the absence of the quantum gravity effect, can be considered as the signature of the quantum gravity on the CMB. The quantum gravity effect on the lensed BB mode correlation angular power spectrum with BK15 and Planck 2018 joint data does not rule out the chaotic inflationary models.

Keywords Quantum gravity · Inflation · CMB

1 Introduction

Despite several approaches developed in understanding quantum gravity over decades, still there is no complete and satisfactory theory of quantum gravity. Therefore, a viable theory for quantum gravity remains a formidable challenging task even today. However, it does not prevent us from exploring the observational consequence of quantum gravity effect through some suitable experiments. Recently, it is remarked that the effect of quantum gravity can be explored through cosmic microwave background radiation (CMB) through inflation and appears to be very promising. Hence, hopefully, the CMB can be used as a testing arena for any theory of quantum gravity. Inflation is a scenario proposed to resolve some of the shortcomings of the standard cosmology [1–8]. According to the inflationary scenario, the universe expanded several folds for a brief interval of time during its very early stage of evolution. One of the attractive features of inflation is that it seeded the formation of large-scale structures in the universe and also possibly generated primordial gravitational waves that are yet to be discovered. Therefore, the detection of the primordial gravitational

P. K. Suresh (✉)
School of Physics, University of Hyderabad, Hyderabad, Telangana, India
e-mail: sureshpk@uohyd.ac.in

K. S. Sreelatha and V. Jacob (eds.), *Modern Perspectives in Theoretical Physics*,
https://doi.org/10.1007/978-981-15-9313-0_3

waves can not only provide ample evidence of its existence but also tells whether the universe had an inflationary era or not. It is believed that the primordial gravitational waves left its imprint on the CMB in the form of B-mode polarization [9, 10]. The B-mode polarization of the CMB is a unique feature due to the primordial gravitational waves. A recent study indicates that the B-mode polarization of the CMB is useful in exploring the effect of quantum gravity through inflation, provided the energy scale of inflation field was much larger than the Planck scale. It is reasonable to assume that such a high energy scale, the field that is responsible for inflation necessarily be sensitive enough to quantum gravity. Therefore, the exploration of the B-mode polarization of the CMB is very important in understanding inflation as well as quantum gravity. Recent results of the B-mode polarization of the CMB [11] with Lyth bound [12] indicate that the value of the inflation field could be much higher than the Planck energy scale. This provides a novel and promising path to explore quantum gravity experimentally through the B-mode polarization of the CMB.

As discussed, there is no consistent theory of quantum gravity existing at present. Nevertheless, currently, several approaches available to it and one such approach is the effective theory [13, 14]. The feasibility of the effective theory techniques to inflation also has already been examined [15, 16]. Further, it is shown that the effective theory scenario has the prospective to accommodate some of the disfavored inflationary models because of the nature of the higher dimensional operator of the theory [13]. As per the effective theory, the higher dimensional operator of inflation potential can take only small value, but can play a key role in the inflation and consequently on the BB mode spectrum of the CMB. The quantum gravity effect may result in reducing the value of the slow-roll parameters and the tensor-to-scalar ratio of an inflationary model compared to its predicted value. The consequence of the lowering of the tensor-to-scalar ratio is expected to reflect on the BB mode angular power spectrum of the CMB. Therefore, it is interesting to explore the imprint of quantum gravity on the BB mode angular power spectrum of the CMB through inflation with the measured B-mode polarization data of the CMB. The repercussion of the lowering value of the tensor-to-scalar ratio of an inflation model on the BB mode angular spectrum of the CMB is a potential probe to seek the signature of quantum gravity. Therefore, the present study aims to explore the signature of the quantum gravity that follows the effective theory approach via the chaotic inflationary models in terms of the BB mode correlation angular power spectrum with the BK15 and Planck 2018 joint data of the CMB [17].

Throughout the study we follow the unit $c = \hbar = 1$.

2 Inflation and Power Spectra

In a simple inflationary scenario, a homogeneous scalar field, known as the inflation, is considered as the candidate of inflaton. The equation of motion of the inflaton, called the Klein–Gordon equation, in a flat Friedmann–Lemaître–Robertson–Walker (FLRW) metric can be written as

$$\ddot{\phi} + 3H\dot{\phi} + V' = 0, \tag{1}$$

where V is the inflaton potential, prime means derivative with respect to the inflaton field and $H = \frac{\dot{a}}{a}$ (a is the scale factor of the FLRW metric) is the Hubble parameter governed by the Friedmann equation,

$$H^2 = \frac{1}{3m_{pl}^2}\left(\frac{1}{2}\dot{\phi}^2 + V(\phi)\right), \tag{2}$$

where m_{pl} is the reduced Planck mass and term inside the bracket is the energy density of the inflaton.

In a wider sense, inflation means a phase of accelerated expansion of the universe. Actually, the dynamical systems corresponding to (1) and (2) do not necessarily lead to an accelerated expansion always; however, the inflation can take place when the potential energy of the inflaton dominates over its kinetic energy, that is, under the slow-roll regime. Hence, under the slow-roll condition, that is, when $\frac{\dot{\phi}^2}{2} \ll V$, (2) takes a simple form:

$$H^2 \simeq \frac{V}{3m_{pl}^2}, \tag{3}$$

and the Klein–Gordon equation (1) becomes

$$3H\dot{\phi} + V' \simeq 0. \tag{4}$$

Therefore, using Eq. (4) with slow-roll condition gives the first slow-roll parameter [18–20]

$$\epsilon_v = \frac{m_{pl}^2}{2}\left(\frac{V'}{V}\right)^2, \tag{5}$$

and when $\ddot{\phi} \ll V'$ with (4) and (2) gives the second slow-roll parameter

$$\eta_v = m_{pl}^2\left(\frac{V''}{V}\right), \tag{6}$$

and so on, with the conditions $\epsilon_v, \eta_v \ll 1$. The inflation terminates when $\epsilon(\phi_{end}) = 1$ but it should last long enough to resolve at least some of the shortcomings of the hot Big Bang model. The duration of inflation is usually characterized by a number known as e-folding number N, defined by

$$N = \ln\frac{a_{end}}{a_{inf}}, \tag{7}$$

where a_{end} and a_{inf} are, respectively, the values of the scale factor at the end of the inflation and during the inflation. According to the definition, N decreases during the inflationary era and becomes zero at the end of the inflation stage. By considering this fact and under the slow-roll condition the e-folding number can be expressed in terms of the inflaton field as

$$N \simeq \frac{1}{m_{pl}} \int_{\phi_{end}}^{\phi} \frac{1}{\sqrt{2\epsilon}} d\phi. \tag{8}$$

Thus, for a given inflation potential, one can estimate N in terms of the inflaton field.

The isotropic and homogeneous feature of cosmology is insufficient in understanding the actual universe, and hence deviation from the isotropy and homogeneity is very much essential. The growth of the inhomogeneities due to the attractive nature of gravity implies that the inhomogeneity was very small in the past. Therefore, the linear perturbation theory is quite adequate to handle most of the evolution of inhomogeneities. Since the linear approximation halt at small scales in the past of the universe, the reconstruction of primordial inhomogeneities from the large-scale structure of the universe has been laborious. However, it is quite sufficient to recount the fluctuations of the CMB at the time of last scattering epoch. Therefore, at present, CMB is the foremost observational inquest of the primordial inhomogeneities.

Next, we consider a perturbed inflation field in a perturbed cosmological geometry. According to Einstein's field equation, the metric fluctuations must necessarily be coexistent with scalar field fluctuations. The linear perturbations can be examined about a flat FLRW background.

Classically, during the inflation the scalar field can be treated as a homogeneous field; however, quantum mechanically there can be still fluctuations due to zero-point oscillations. Therefore, the scalar field can be written as homogeneous part and quantum fluctuations (perturbation) part as follows:

$$\phi(x, t) = \phi(t) + \delta\phi(x, t). \tag{9}$$

The quantum fluctuation is known as scalar fluctuation. The perturbed field $\delta\phi(x, t)$ also satisfies the Klein–Gordon equation. The quantum fluctuations can be characterized by the root mean square of the perturbations known as power spectrum. Therefore, the primordial spectrum of scalar cosmological perturbations generated from vacuum fluctuations during the slow-roll inflation era known as the scalar power spectrum P_S can be written in terms of the inflation potential as [21, 22]

$$P_S = \frac{1}{12\pi^2 m_{pl}^6} \frac{V^3}{V'^2}|_{k=aH}, \tag{10}$$

where $k = aH$ means the wave number crosses the horizon.

In addition to the scalar perturbation, the inflation also generated the tensor perturbations or primordial gravitational waves from the zero-point vacuum fluctuations. The root mean square of the cosmological tensor perturbations can be characterized by the tensor power spectrum and can be written in terms of the inflation potential as follows:

$$P_T = \frac{2}{3\pi^2 m_{pl}^4} V|_{k=aH}. \tag{11}$$

It can be observed that the tensor and scalar power spectra are not independent of the first slow-roll parameter. Therefore, the tensor-to-scalar ratio can be written as

$$r \equiv \frac{P_T(k)}{P_S(k)} = 16\epsilon. \tag{12}$$

This ratio means the amplitude of the tensor perturbation at the CMB scale.

In reality, the tensor and scalar spectra are not necessarily scale invariant because the scalar field varies slowly during inflation. The variation is usually accounted with a quantity known as the spectral index and can be defined, respectively, for the scalar and tensor power spectrum as follows:

$$n_s - 1 = \frac{d \ln P_S}{d \ln k} \tag{13}$$

$$n_t = \frac{d \ln P_T}{d \ln k}, \tag{14}$$

where n_s is known as the scalar spectral index and n_t is the tensor spectral index. The spectral indices can also be written in terms of the slow-roll parameters, hence

$$n_s = 1 + 2\eta_v - 6\epsilon_v \tag{15}$$

and

$$n_t = -2\epsilon_v, \tag{16}$$

therefore comparing with (12) leads to the consistency relation

$$r = -8n_t, \tag{17}$$

Which is, in principle, purely related to the observable quantities.

In the present work, we mainly need only the tensor power spectrum to examine the imprint of quantum gravity on the BB mode angular power spectrum of the CMB.

3 Tensor Power Spectrum

To study the quantum gravity effect on the BB mode spectrum of the CMB, knowledge of the tensor power spectrum is essential. Therefore, to compute the tensor power spectrum we begin with the following perturbed flat FLRW metric in the conformal time (defined as $d\tau = \frac{dt}{a}$), given by

$$ds^2 = a^2(\tau)\left[-d\tau^2 + (\delta_{ij} + h_{ij})dx^i dx^j\right], \tag{18}$$

where δ_{ij} means the flat space metric and h_{ij} is the tensor perturbation with the conditions $|h_{ij}| \ll \delta_{ij}$, $\partial_i h^{ij} = 0$, and $\delta^{ij} h_{ij} = 0$.

The tensor perturbation field $h_{ij}(\mathbf{x}, \tau)$ can be written in Fourier mode

$$h_{ij}(\mathbf{x}, \tau) = \frac{\mathcal{B}}{(2\pi)^{\frac{3}{2}}} \int_{-\infty}^{+\infty} \frac{d^3\mathbf{k}}{\sqrt{2k}} \sum_{p=+,\times} \left[h_k^{(p)}(\tau) b_k^{(p)} e^{i\mathbf{k}.\mathbf{x}} \chi_{ij}^{(p)}(\mathbf{k})\right.$$
$$\left. + h_k^{(p)*}(\tau) b_k^{(p)\dagger} e^{-i\mathbf{k}.\mathbf{x}} \chi_{ij}^{(p)*}(\mathbf{k})\right], \tag{19}$$

where $\mathcal{B} = \sqrt{16\pi G}$, $\mathbf{k}$ is the wave vector, and $\chi_{ij}^{(p)}$ (where $p = +, \times$) are the two linear polarization states of gravitational wave that satisfy $\chi_{ij}^{(p)}\delta^{ij} = 0$, $\chi_{ij}^{(p)}k^i = 0$, $\chi_{ij}^{(p)}\chi^{(p')ij} = 2\delta_{pp'}$, $\chi_{ij}^{(p)}(\text{-}\mathbf{k}) = \chi_{ij}^{(p)}(\mathbf{k})$, and are, respectively, called plus $(+)$ polarization and cross $(\times)$ polarization. In (19), $(b_k^{(p)\dagger})$ and $(b_k^{(p)})$ are, respectively, the creation and annihilation operators and are governed by the Heisenberg equation with Hamiltonian $\mathcal{H}$ for each component, given by

$$\frac{d}{d\tau} b_k^{(p)\dagger}(\tau) = -i\left[b_k^{(p)\dagger}(\tau), \mathcal{H}\right], \quad \frac{d}{d\tau} b_k^{(p)}(\tau) = -i\left[b_k^{(p)}(\tau), \mathcal{H}\right], \tag{20}$$

and further the operators satisfy the conditions $[b_k^{(p)}, b_{k'}^{(p')\dagger}] = \delta_{pp'}\delta^3(k - k')$ and $[b_k^{(p)}, b_{k'}^{(p')}] = [b_k^{(p)\dagger}, b_{k'}^{(p')\dagger}] = 0$.

The initial vacuum state can be defined with respect to the annihilation operator as

$$b_k^{(p)}|0\rangle = 0.$$

Since the contribution from each polarization to the gravitational wave is the same, hereafter we drop the superscript (p) for convenience.

The value of the operators $b_k^\dagger(0)$ and $b_k(0)$ at some initial time, say $t = 0$, to its later time, say $t = \tau$, can be connected through the Bogoliubov transformation given by

$$b_k^\dagger(\tau) = u_k^*(\tau) b_k^\dagger(0) + v_k^*(\tau) b_k(0), \tag{21}$$
$$b_k(\tau) = u_k(\tau) b_k(0) + v_k(\tau) b_k^\dagger(0), \tag{22}$$

where the complex functions $u_k(\tau)$ and $v_k(\tau)$ hold the condition $|u_k|^2 - |v_k|^2 = 1$.

The coupling of the tensor perturbation $h_k(\tau)$ with the scale factor $a(\tau)$ gives

$$h_k = \frac{\xi_k}{a}, \tag{23}$$

where

$$\xi_k(\tau) = u_k(\tau) + v_k^*(\tau) \tag{24}$$

satisfies the equation of motion

$$\xi_k'' + \left(k^2 - \frac{a''}{a}\right)\xi_k = 0. \tag{25}$$

Here prime means derivative with respect to the conformal time τ.

Since the contribution from each polarization is the same, using (19) and (23), we get

$$h(\mathbf{x}, \tau) = \frac{\mathcal{B}}{a(\tau)(2\pi)^{\frac{3}{2}}} \int_{-\infty}^{+\infty} d^3\mathbf{k}[\xi_k(\tau)b_k + \xi_k^*(\tau)b_k^\dagger]e^{i\mathbf{k}.\mathbf{x}}. \tag{26}$$

The two-point correlation function of the tensor perturbation in a vacuum state is given by

$$\langle h_k h_{k'}^* \rangle = \frac{2\pi^2}{k^3} P_T(k)\delta^3(\mathbf{k} - \mathbf{k}'), \tag{27}$$

where P_T is the tensor power spectrum in the super horizon limit ($k << aH$).

Therefore, using (19) and (27), we get the primordial tensor power spectrum and it can be expressed in a power law form as

$$P_T(k) = A_T(k_0)\left(\frac{k}{k_0}\right)^{n_t}, \tag{28}$$

where $A_T(k_0) = \mathcal{B}^2\left(\frac{H}{2\pi}\right)^2$ is the amplitude at the pivot wavenumber $k_0 = aH$ and H is the Hubble parameter during the inflation.

4 Effective Theory and Inflation

The effective theory is an approach to quantum gravity. Therefore, it is interesting to study the inflation in the light of the effective theory and its reflection on the BB mode angular power spectrum of the CMB. The action for the scalar field coupled to gravity in the most general effective theory is [13]

$$S = \int d^D \sqrt{-g} \left(\frac{m_{pl}^2}{2} R + f(\phi) F(R, R_{\mu\nu}) + g^{\mu\nu} \partial_\mu \phi \partial_\nu \phi \right.$$

$$\left. + V_{ren}(\phi) + \sum_{n=5}^{\infty} c_n \frac{\phi^n}{m_{pl}^{n-4}} \right), \tag{29}$$

where $V_{ren}(\phi)$ includes all renormalizable terms up to dimension four and c_n are Wilson coefficients of higher dimensional operators [13]. The second term in the bracket corresponds to the non-minimal coupling between the gravity and the scalar field. The present study assumes that the coupling of the scalar field to gravity is minimal, and hence ignoring the second term. We also neglect the derivative term like $\partial_\mu \phi \partial^\nu \phi^n$ for simplicity. Further, it is assumed that the kinetic term of the graviton is canonically normalizable.

In view of the effective theory approach to inflation, the potential for the inflaton can be written as

$$V(\phi) = V_{ren}(\phi) + \sum_{n=5}^{\infty} c_n \frac{\phi^n}{m_{pl}^{n-4}}. \tag{30}$$

The term $c_n \frac{\phi^n}{m_{pl}^{n-4}}$ present in the above potential is known as the higher dimension operator (HDO). For a given inflationary model, usually one specific dominant term of Wilson coefficients is only considered and other remaining terms are ignored. To preserve the flatness of the inflationary potential, the value of Wilson coefficients must be of the order of 10^{-3} [4, 5, 23]. The origin of HDO depends on the nature of the quantum gravity theory. For instance, in the extra-dimensional scenario it may arise from the exchange of supermassive Kaluza–Klein mode or due to either real or virtual quantum black holes in generic quantum gravity theory [24].

4.1 *Quantum Gravity Effect on the Chaotic Inflation*

Next, we consider the effect of quantum gravity on the two specific chaotic inflationary models, namely, quadratic and quartic chaotic inflation models through the HDO effect. In the usual scenario, these inflationary models are disfavorable due to the Planck result [25–28]. However, the study of the inflation model with the HDO effect is encouraging to reconsider some of the disfavored models because it may bring down the value of the associated parameters of the inflation compared to their standard case.

4.1.1 Quadratic Chaotic Inflation

In the light of the effective theory, the potential for the quadratic chaotic inflation can be written as [13]

$$V(\varphi) = m_{pl}^4 \left(\bar{m}^2\varphi^2 + c_n\varphi^n\right), \tag{31}$$

where $\bar{m} = m/m_{pl}$ and $\varphi = \phi/m_{pl}$ are, respectively, normalized mass and the normalized inflation field with the reduced Planck mass. The first term in (31) is the potential for the quadratic chaotic inflation in the absence of the HDO.

In principle, it is possible to expand the HDO, but for the present work, we consider only the dominating term as c_6, and others are set to zero because their effect is subdominant. As noted in [13], the HDO should be a correction to the leading term on $\bar{m}^2$. Therefore, $c_6 = \alpha_m\bar{m}^2$ and it implies that the potential (31) can be recasted as

$$V(\varphi) = m_{pl}^4\bar{m}^2\varphi^2 \left(1 + \alpha_m\varphi^4\right), \tag{32}$$

with the following condition:

$$\mid \alpha_m \mid \varphi^4 < 1, \tag{33}$$

the potential (32) can be expanded.

The first slow-roll parameter corresponding to the quadratic chaotic inflation potential with HOD effect can be computed using (5) and (32), and therefore by neglecting higher order terms, we get

$$\epsilon_V = \frac{2m_{pl}^2}{\varphi^2}\left(1 + 4\alpha_m\varphi^4\right). \tag{34}$$

Similarly the second slow-roll parameter of the potential is obtained using (6) and (32), and then by neglecting higher order terms, we get

$$\eta_V = \frac{2m_{pl}^2}{\varphi^2}\left(1 + 14\alpha_m\varphi^4\right). \tag{35}$$

Therefore, tensor-to-scalar ratio of the quadratic chaotic inflation potential with HDO effect is

$$r = \frac{32m_{pl}^2}{\varphi^2}\left(1 + 4\alpha_m\varphi^4\right). \tag{36}$$

Inflation terminates when the first slow-roll parameter becomes unity, and therefore using (8) the inflation field can be written in terms of the e-folding number as

$$\varphi_N^2 \simeq \frac{N}{2\pi}\left(1 + \frac{N^2\alpha_m}{6\pi^2}\right). \tag{37}$$

The initial value of the inflaton at the beginning stage of inflation depends on $\alpha_m N^2$ but (33) implies that

$$\mid \alpha_m \mid \varphi_N^4 \simeq \frac{N^2 \mid \alpha_m \mid}{4\pi^2} \lesssim 1. \tag{38}$$

It is possible to estimate an effective field theoretical bound of the HDO by considering a large e-folding number and is

$$\mid \alpha_m \mid^{EFT} \lesssim 2 \times 10^{-2}. \tag{39}$$

The first slow-roll parameter of the quadratic inflation with HDO effect can be obtained in terms of e-folding number by (34) and (37) as

$$\epsilon_V = \frac{1}{2N}\left(1 + \frac{5}{6}\frac{N^2\alpha_m}{\pi^2}\right), \tag{40}$$

hence the obtained tensor-to-scalar ratio of quadratic inflation with HDO effect is

$$r = \frac{8}{N}\left(1 + \frac{5}{6}\frac{N^2\alpha_m}{\pi^2}\right). \tag{41}$$

4.1.2 Quartic Chaotic Inflation

By considering the effective theory approach, the potential for the quartic chaotic inflation can be written as [13]

$$V(\varphi) = m_{pl}^4 \left(\lambda\varphi^4 + c_n\varphi^n\right), \tag{42}$$

where $\varphi = \phi/m_{pl}$ is the normalized inflation field for the quartic inflation with the reduced Planck mass. The first term in (42) is the potential corresponding to the standard quartic chaotic inflation model.

Here also we consider only the dominant term as c_6 in (42), and hence the HDO correction is taken as the leading term on λ as $c_6 = \alpha_\lambda\lambda$, which implies that the quartic chaotic inflation with HDO effect is

$$V(\varphi) = m_{pl}^4\lambda\varphi^4\left(1 + \alpha_\lambda\varphi^2\right), \tag{43}$$

with the condition

$$\mid \alpha_\lambda \mid \varphi^2 < 1, \tag{44}$$

hence the quartic chaotic inflation potential (43) with HDO effect can be expanded.

By using (5) and (43) the first slow-roll parameter for the quartic chaotic inflation with HDO effect is obtained by ignoring the higher order terms as

$$\epsilon_V = \frac{8m_{pl}^2}{\varphi^2}\left(1+\alpha_\lambda\varphi^2\right), \tag{45}$$

and (6) and (43) give the second slow-roll parameter for the quartic chaotic inflation with HDO effect by neglecting the higher order terms as

$$\eta_V = \frac{4m_{pl}^2}{\varphi^2}\left(1+\frac{5}{2}\alpha_\lambda\varphi^2\right). \tag{46}$$

Using the condition to end of inflation in terms of the slow-roll parameter and using (8), (45) we can express the inflation field in terms of the e-folding number as

$$\varphi_N^2 \simeq \frac{N}{\pi}\left(1+\frac{N\alpha_\lambda}{4\pi}\right). \tag{47}$$

The initial value of the inflaton at the beginning stage of inflation depends on $\alpha_\lambda N$; however, (44) implies that

$$\mid \alpha_\lambda \mid \varphi_N^2 \simeq \frac{N \mid \alpha_\lambda \mid}{\pi} \lesssim 1. \tag{48}$$

An estimate of HOD for the quartic chaotic inflation with HDO effect with a large e-folding number gives the effective field theoretical bound of HDO as

$$\mid \alpha_\lambda \mid^{EFT} \lesssim 0.06. \tag{49}$$

Using (45) and (47), the tensor-to-scalar ratio for the quartic chaotic inflation with HDO effect is obtained in terms of the e-folding number as

$$r = \frac{16}{N}\left(1+\frac{3}{4}\frac{N\alpha_\lambda}{\pi}\right). \tag{50}$$

Therefore, η, n_s, and n_t for the quartic chaotic inflation with HDO effect can also be obtained in terms of e-folding number N.

We can compute the value of the tensor-to-scalar ratio for the chaotic inflationary models for various values of the HDO parameter. The estimated value of the tensor-to-scalar ratio for the quadratic and quartic chaotic inflationary models for various values of the HDO parameter is presented in Table 1. It can be observed that the quantum gravity effect on inflation is reducing the value of the tensor-to-scalar ratio for both quadratic and quartic inflation models compared to their standard case. Note that the recent estimate of the upper bound of the tensor-to-scalar ratio from the CMB observation is $r_{0.002} < 0.064$ [29]. Hence, it may be concluded that the quadratic and

Table 1 Estimated value of the tensor-to-scalar ratio for the quadratic and quartic inflation models for various values of their respective HDO parameter

Quadratic		Quartic	
α_m	r	α_λ	r
0	0.133	0	0.267
−0.0015	0.0725	−0.050	0.0756
−0.0017	0.0644	−0.053	0.0641
−0.0019	0.0563	−0.061	0.0336
−0.0021	0.0481	−0.072	0.0084

quartic chaotic inflationary models with quantum gravity cannot be ruled out with the current estimated tensor-to-scalar ratio obtained from various CMB observations.

Further, note that most of the values of the HDO parameter of the chaotic inflation model presented in the table are less than the upper limit of the effective field theory estimate.

5 Quantum Gravity Effect on the BB Mode Angular Spectrum of CMB

In this section, we study the quantum gravity effect on the BB mode angular power spectrum of the CMB through the chaotic inflationary models with BK15 and Planck 2018 joint data.

The primordial gravitational waves are responsible for the B-mode polarization of the CMB [30, 31]. The resulting BB mode correlation angular power spectrum can be computed with following expression [32, 33]:

$$C_l^{BB} = (4\pi)^2 \int dk\, k^2\, P_T(k) \times \left| \int_0^{\tau_0} d\tau g(\tau) h_k(\tau) \left[2 j_l'(x) + \frac{4 j_l(x)}{x} \right] \right|^2,$$

where $g(\tau) = \kappa' e^{-\kappa}$ is the visibility function and κ' is the differential optical depth, j_l is the spherical Bessel function, and $x = k(\tau_0 - \tau)$.

The unlensed and lensed BB mode correlation angular spectrum of the CMB with quantum gravity effect for the quadratic and quartic chaotic inflation is obtained. For this, the optical depth is taken as $\kappa = 0.08$, the tensor pivot wave number is taken as $k_0 = 0.002\ \mathrm{Mpc}^{-1}$, and the scalar pivot is taken as $k_0 = 0.002\ \mathrm{Mpc}^{-1}$. The obtained spectra are normalized with COBE.

The unlensed BB mode correlation angular spectrum for various values of the HDO parameter for the quadratic and quartic inflation, respectively, is shown in

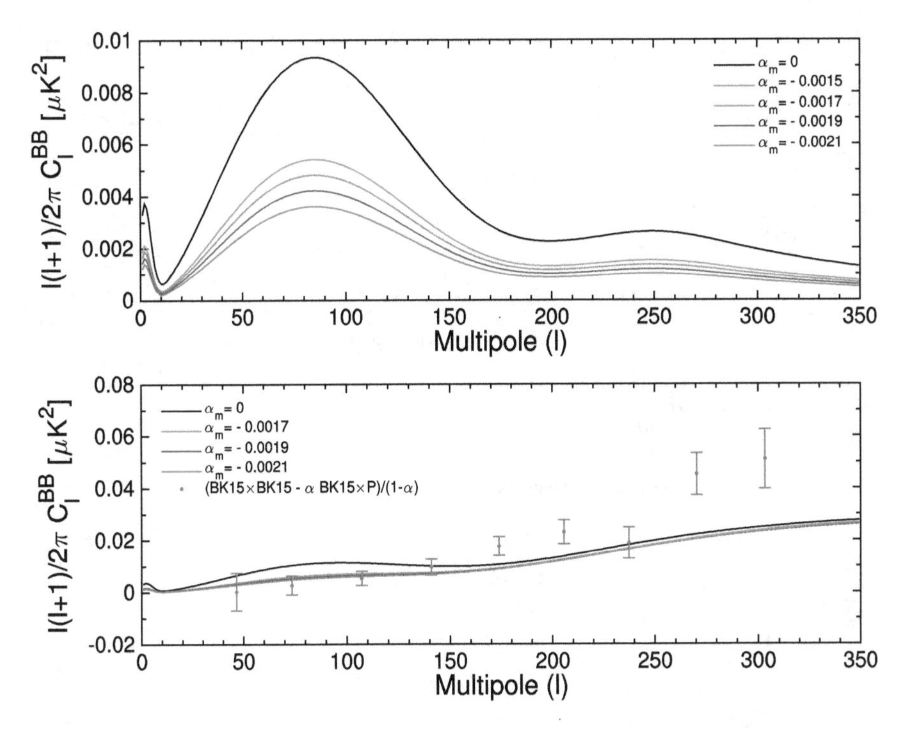

Fig. 1 Quantum gravity effect (solid color lines) on the unlensed (upper panel) BB mode angular spectrum of CMB for the quadratic chaotic inflation and its lensed angular spectrum (lower panel) for various values of HDO with joint BK15 150 GHz and Planck 2018 353 GHz data

the upper panel of Figs. 1 and 2. They show the effect of quantum gravity on the unlensed BB mode spectrum of the CMB. It can be noticed that the power level of the BB mode spectrum is lower than the zero quantum gravity effect. In the absence of quantum gravity effect, the angular power spectrum reduces to the standard case of both inflationary models.

The computed lensed BB mode angular power spectrum of the CMB for various values of HDO, with BK15 and Planck 2018 joint data analysis, for the quadratic and quartic inflation, is, respectively, shown in the lower panel of Figs. 1 and 2. The observed B-mode polarization data contains some dust contribution. Therefore, to remove the dust contribution for the combination (BK15 $\times$ BK15 $-\alpha$ BK15 $\times$ P)/(1 $-$ α) with the autocorrelation spectrum of BK15 at 150 GHz map and cross-correlation spectrum of BK15 at 150 GHz map with Planck 2018 at 353 GHz map, we adopt the same model that implemented for the combination (BK $\times$ BK $-\alpha$ BK $\times$ P)/(1 $-$ α) for the auto spectrum of BK at 150 GHz map and cross-spectrum of BK at 150 GHz map and Planck at 353 GHz map [34]. In the dust removal model, it is assumed that the dust contribution model parameter $\alpha = \alpha_{fid} = 0.04$, and is the implemented value in the case of the BICEP2/Keck Array 150 GHz and Planck 353 GHz combination [34].

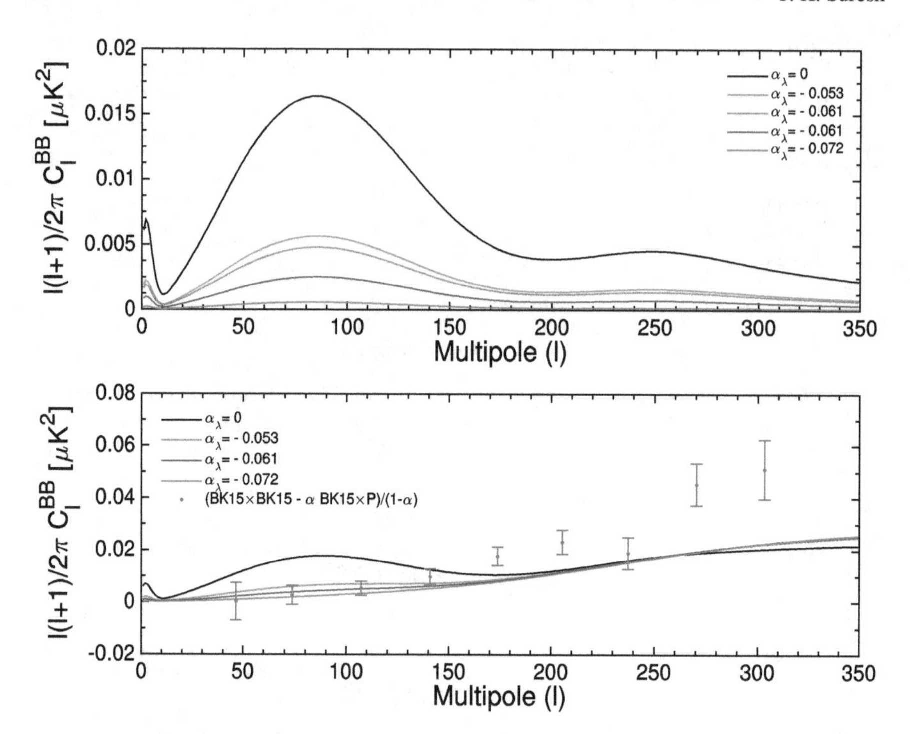

Fig. 2 Quantum gravity effect (solid color lines) on the unlensed (upper panel) BB mode angular spectrum of CMB for the quartic chaotic inflation and its lensed angular spectrum (lower panel) for various values of HDO with joint BK15 150 GHz and Planck 2018 353 GHz data

The lensed BB mode angular spectrum corresponding to the quantum gravity effect for the quadratic inflation lies marginally within the limit of BK15 and Planck 2018 joint data for $l < 150$, but at the same time out of the limit for most of the other higher multipole moments. The lensed angular spectrum for the quantum gravity effect is out of the limit of BK15 and Planck 2018 joint data for most of the multipoles.

The obtained lensed BB mode angular spectrum for the quartic inflation is compared with the BK15 and Planck 2018 joint data. The lensed angular spectrum corresponding to the quantum gravity effect for the quartic case lies within the limit of BK15 and Planck 2018 joint data for $l < 150$, but out of the limits for most of the other higher multipole moments.

From the study, it can be observed that the lensed BB mode spectrum of CMB for the higher multipole moments is out of the limit of the BK15 and Planck 2018 joint data for both inflationary models and it appears that quantum gravity effect cannot be fully accounted for its compatibility with observed results. On the other hand, the lower multipole region of the BB mode spectrum lies within the observed limit of the BK15 and Planck 2018 joint data. Therefore, the reduction of the power of the BB mode spectrum of the CMB, especially for the lower multipole less than 150 of the lensed spectrum, can be considered as the signature of the quantum gravity. Further, from the analysis of the study, it may be concluded that the quantum gravity effect on the lensed BB mode correlation angular power spectrum of CMB with BK15 and Planck 2018 joint data shows the quadratic and quartic chaotic inflationary models cannot be ruled out at present.

6 Conclusion

The realization of the quantum nature of gravity both theoretically and experimentally is an equally formidable challenging task. Though several approaches to quantum gravity were proposed over the years, there is still a problem of lack of a consistent theory. However, this does not prevent to explore the quantum gravity employing appropriate experiment or else testing its consequence on the other phenomena. One of the potential candidates to realize quantum gravity experimentally is the CMB via inflation, provided presumably the energy scale of the inflation field possessed higher value than the Planck scale. In such a higher energy scale, the inflation field is assumed to be sensitive enough to quantum gravity. The recent result of the CMB observation with Lyth bound suggests that the scale of the inflation field, in principle, can take higher value than its standard case. Recently, it has been suggested that the effective theory approach to quantum gravity and hence its effect on inflation is very much suitable to observe the consequence of quantum gravity experimentally. The effective theory approach to quantum gravity is one of the alternative approaches to quantum gravity. Given the effective theory scenario, inflationary potential under the consideration with the HDO parameter can have some impact on slow-roll parameters of the inflation, and hence its consequence is expected to reflect on the BB mode correlation angular spectrum of the CMB.

Based on the effective theory approach it is proposed that some of the disfavored inflationary models may be reconsidered if the quantum gravity had played a key role in the dynamics of the inflation. The extension of the effective theory approach to the chaotic inflation showed that the tensor-to-scalar ratio of the model is lower than its predicted value. The most effective tool to test any inflation model is to examine the B-mode polarization of CMB due to the primordial gravitational waves generated during inflation. Therefore, the BB mode angular power spectrum of CMB can be used to examine inflation as well as quantum gravity. In the present work, we investigated the quantum gravity effect on the BB mode correlation angular power spectrum of CMB through the quadratic and quartic inflation models with BK15 and Planck 2018 joint data.

We examined the effect of quantum gravity on the lensed BB mode correlation angular power spectrum of CMB for the quadratic and quartic inflation for the combination (BK15 $\times$ BK15 $-\alpha$ BK15 $\times$ P)/(1 $-$ α) of the autocorrelation spectrum of BK15 at 150 GHz map and cross-correlation spectrum BK15 at 150 GHz map and Planck 2018 at 353 GHz map with the cleaned dust contribution. From the study, it can be observed that the lensed BB mode spectrum of CMB for the higher multipole moments is out of the limit of the BK15 and Planck 2018 joint data for both inflationary models and it appears that quantum gravity effect cannot be fully accounted for its compatibility with observed results. On the other hand, the lower multipole region of the BB mode spectrum lies within the observed limit of the BK15 and Planck 2018 joint data. Therefore, it may be concluded that the reduction of the power of the BB mode spectrum of the CMB, especially for the lower multipole less than 150, can be considered as the signature of the quantum gravity. Further, from the analysis of the quantum gravity effect on the lensed BB mode correlation angular power spectrum of CMB with BK15 and Planck 2018 joint data, it may be concluded that the chaotic inflationary models cannot be ruled out.

The values of HDO parameter relevant to each inflation model considered in the present study are constrained with tensor-to-scalar ratio estimated from various current observed results of the CMB. However, the tensor-to-scalar ratio may be further restricted to the estimate of the upcoming CMB observations such as PICO, CMB-S4, etc. Therefore, the value of the HDO parameter may be constrained further with the upcoming missions of CMB and may be useful to examine various inflationary models. Thus, the results of the present study may be useful in validating the inflationary model and hopefully quantum gravity also.

In the present work, we examined the signature of quantum gravity on the BB angular power spectrum through the chaotic inflationary models only. However, one can re-examine the effect of quantum gravity with other inflationary models as well as with the other theories of quantum gravity which is beyond the scope of the present work.

Acknowledgements P K S acknowledges the financial support of DST-SERB, New Delhi. Author would like to thank Dr. Rizwan Ul Haq Ansari for the wonderful discussions. Author acknowledges the use of CAMB code and thanks Planck and BICEP team and websites for the data.

References

1. A.H. Guth, Phys. Rev. D **23**, 347 (1981)
2. R.H. Brandenberger, IPM School on Cosmology 1999. arXiv:hep-ph/9910410v1 (1999)
3. A.D. Linde, *Particle Physics and Inflationary Cosmology* (CRC Press, 1990)
4. A.D. Linde, Lect. Notes Phys. **738**, 1 (2008)
5. A.D. Linde, Phys. Letts. B **129**, 177 (1983)
6. A.H. Guth, S.Y. Pi, Phys. Rev. Lett. **49**, 1110 (1982)
7. A.D. Linde, Phys. Rev. D **49**, 748 (1994)
8. J. Garcia-Bellido, A.D. Linde, Phys. Rev. D **57**, 6075 (1998)
9. S. Dodelson, *Modern Cosmology* (Academic Press, 2003)
10. S. Dodelson, W.H. Kinney, E.W. Kolb, Phys. Rev. D **56**, 3207 (1997)
11. P.A.R. Ade et al., Phys. Rev. Lett. **112**, 241101 (2014)
12. D.H. Lyth, Phys. Rev. Lett. **78**, 1861 (1997)
13. X. Calmet, V. Sanz, Phys. Lett. B **737**, 12 (2014)
14. J.F. Donoghue, AIP. Conf. Proc. **1483**, 73 (2012)
15. N. Agarwal, R.H. Ribeiro, R. Holman, JCAP **06**, 016 (2014)
16. H. Collins, R. Holman, T. Vardanyan. arXiv:1403.4592 [hep-th]
17. P.A.R. Ade et al., Phys. Rev. Lett. **121**, 221301 (2018)

18. D.J. Schwarz, C.A. Terrero-Escalante, A.A. Garcia, Phys. Lett. B **517**, 243 (2001)
19. S.M. Leach, A.R. Liddle, J. Martin, D.J. Schwarz, Phys. Rev. D **66**, 023515 (2002)
20. D.J. Schwarz, C.A. Terrero-Escalante, JCAP **0408**, 003 (2004)
21. S. Kuroyanagi, T. Takahashi, JCAP **1110**, 006 (2011)
22. S. Kuroyanagi, T. Chiba, N. Sugiyama, Phys. Rev. D **79**, 103501 (2009)
23. R. Holman, S.D.H, Hsu, T.W. Kephart, E.W. Kolb, R. Watkins, L.M. Widrow, Phys. Lett. B **282**, 132 (1992)
24. X. Calmet, D. Fragkakis, N. Gausmannh, Eur. Phys. J. C **71**, 1781 (2011)
25. P.A.R. Ade et al., Astron. Astrophys. **571**, A22 (2014)
26. J. Martin, C. Ringeval, V. Vennin, arXiv:1303.3787 (2013)
27. J. Martin, C. Ringeval, R. Trotta, V. Vennin, JCAP **03**, 039 (2014)
28. J. Martin, arXiv:1502.05733v1 (2015)
29. Y. Akrami et al., Planck Collobration. arXiv:1807.06211
30. A. Kosowsky, New Astron. Rev. **43**, 157 (1999)
31. Y.T. Lin, B.D. Wandelt, Astropart. Phys. **25**, 151 (2006)
32. U. Seljak, M. Zaldarriaga, Phys. Rev. Lett. **78**, 2054 (1997)
33. D. Baskaran, L.P. Grishchuk, A.G. Polnarev, Phys. Rev. D **74**, 083008 (2006)
34. P.A.R. Ade et al., Phys. Rev. Lett. **114**, 101301 (2015)

High Energy Physics and Gravitation

Aspects of Schrödinger Picture Formalism

K. P. Satheesh

Abstract Application of functional Schrödinger picture formalism to scalar theories is described at an elementary level. The basic formalism is applied to a free scalar field. To handle interacting fields, variational approximation with Gaussian wave functional as trial wave functional is developed. The procedure is applied to ϕ^4 and ϕ^6 self-interacting scalar fields. The effective action, effective potential and effective mass are calculated up to two-loop and three-loop levels for ϕ^4 and ϕ^6, respectively. Calculation of static effective potential is done and compared with Gaussian effective potential formalism and CJT formalism.

Keywords Schrödinger picture · Quantum field theory · De-coherence · Gaussian effective potential

1 Introduction

Schrödinger picture formalism in quantum field theory is a generalization from ordinary quantum mechanics to the infinite number of degrees of freedom which comprises a field. Quantum state in this approach is a functional superposition of field configurations. In studying the detailed structures of both bosonic and fermionic quantum fields, the variational approximation applied to functional Schrödinger picture has been found to be very useful. Here I use the simplest variational approximation which is a Gaussian approximation and is found to provide better information than large-N approximation. It is found that the effective potential expression obtained using Schrödinger picture and Cornwall–Jackiw–Tomboulis formalism opens a channel towards finite temperature field theories. Vacuum properties in this formalism are conceptually simpler than Fock Space vacuum (normal ordering not required) and find several applications in quantum fields in curved space-time.

K. P. Satheesh (✉)
Government Brennen College, Thalassery, Kerala, India
e-mail: satheeshsithara@gmail.com

Institute for Intensive Research in Basic Sciences, M.G University, Kottayam, Kerala, India

K. S. Sreelatha and V. Jacob (eds.), *Modern Perspectives in Theoretical Physics*,
https://doi.org/10.1007/978-981-15-9313-0_4

Functional Schrödinger picture was found on the fact that quantum states and operators can be projected on the field amplitude basis. Historically the quantum mechanics put forward by Schrödinger is immensely successful in its predictions about the behaviour of particles at quantum level. But for quantizing classical and quantum fields we conventionally do not use Schrödinger equation explicitly. In this review, I treat quantum theory of fields by extending the concepts and methods of Schrödinger approach to quantum mechanics [1]. In quantum mechanics, in Schrödinger picture, the time dependence of observables is encoded in the states while the operators remain time independent. In quantum field theory using Schrödinger picture, this implies a time-independent field operator $\dot{\hat{\phi}} = 0$. Canonical quantization is implemented by demanding commutation relations for conjugate operators. In the coordinate representation, position operators are represented by their eigenvalues and momentum operators by differential equations and states by wave functions. In field theory, we use field operators instead of position operators whose eigenvalues are functions instead of numbers. The states are represented by wave functional and Schrödinger equations become functional differential equations.

The techniques which are already tested in quantum mechanics can be used here. Schrödinger equation describes the time evolution of wave functions which are the coordinate representations of state vectors. Time-dependent Schrödinger equation determines their unitary time evolution. Dynamical variables are treated as Hermitian operators.

In conventional QFT, both fields and their conjugate momenta are treated as operators and used to define the Hamiltonian of the field through a Legendre transformation of the Lagrangian. In the present treatment, fields are treated as coordinate functional and Schrödinger equation finds an extension to obtain the functional Schrödinger equation. Thus, QFT is basically a solution of functional Schrödinger equation.

An exact solution is usually obtained when we treat a free field but approximations are always required when we handle interacting fields. One of the simplest approximations popularly used in quantum mechanics is the variational approximation where a trial state is expressed in terms of certain parameters which can be later extremized to obtain bounds of relevant eigenvalues.

In functional Schrödinger picture also I employ an extension of this procedure which is a Gaussian trial wave functional with variational parameters to implement variational approximation. Here the study is restricted to scalar fields only but can be extended to fermionic and gauge fields with slight modifications. Recently, great interest in the Schrödinger picture formalism of QFT is generated in analysing QFT in curved space-time. Nice features of the vacuum state (vacuum wave functional) in this approach avoid several difficulties associated with QFT in curved space-time [1–4]. The method developed can be comfortably extended to finite temperature QFT [5]. This aspect is not discussed in this review.

The method discussed can be extended to study the de-coherence of massive fields during inflation using variance of non-Gaussian exponent in the wave functional. It is also used to analyse quantum-to-classical transition due to quantum de-coherence [6, 7]. Schrödinger picture is very well suited to explore properties of the vacuum state in non-perturbative regime. The method has been applied to Yang–Mills wave functional in (2 + 1) dimensions [8]. The relation of pre-canonical quantization of gravity to the functional Schrödinger picture is well studied [9]. Guth and Pi, one of the pioneers in the use of Schrödinger picture in QFT, have used it in the study of inflationary perturbations and related the width of the Gaussian vacuum wave functional in de-sitter space-time to Heisenberg picture scalar fields [10].

Analogous to any quantum mechanical theory (for example, path integral formalism) in functional Schrödinger picture also a kernel is a quantum mechanical evolution operator. It represents a transition amplitude from an initial to final state. This has been used in studying the time evolution of vacuum wave functional for QFT on S-brane [11]. Apart from quantum field theory, the procedure is suitable for other many-body systems in condensed matter [12] and also in (1 + 1)-dimensional nonlinear sigma model [13]. The normalized equations of motion for matter and semi-classical gravity in an inhomogeneous space-time have been obtained and Gaussian approximation in functional Schrödinger picture has been used to analyse the time evolution of $\lambda\phi^4$ model [3].

2 Scalar Field Quantization

Quantum field theories generally rely upon the concept of an action defined as $S = \int d^{d+1}\mathcal{L}$ where $\mathcal{L}$ is the Lagrangian density and d is the number of spatial dimensions. The Lagrangian density for the scalar field $\phi(x)$ takes the form

$$\mathcal{L} = \frac{1}{2}\partial_\mu\phi(x)\partial^\mu\phi(x) - V[\phi(x)],$$

where the first term represents the kinetic and the second potential terms. It is worth noting that kinetic term has a larger invariance group than the potential since it is invariant under the shift of the field by a global constant. $\phi \to \phi + a$. Massive scalar field Lagrangian density with mass m is clearly

$$\mathcal{L}_0 = \frac{1}{2}\partial_\mu\phi(x)\partial^\mu\phi(x) - \frac{1}{2}m^2\phi^2.$$

This simple Lagrangian is symmetric under the discrete symmetry $\phi(x) \to -\phi(x)$. If we introduce a quartic self-interacting term to obtain $\mathcal{L}_1 = \mathcal{L}_0 - \frac{\lambda}{4!}\phi^4$ which leads to an acceptable quantum field theory having higher order interactions like $\mathcal{L}_2 = \mathcal{L}_1 - \frac{\xi}{6!}\phi^6$ do not lead to acceptable theories, in general, but gives satisfactory theories in lower dimensions. We can define the canonical momentum operator

$$\Pi(x) = \frac{\partial\mathcal{L}}{\partial(\partial_0\phi)} = \partial_0\phi(x).$$

As usual Hamiltonian is constructed by the Legendre transform of the Lagrangian

$$H(\pi, \phi) = \int d^d x[\pi\dot{\phi} - \mathcal{L}] = \frac{1}{2}\int d^d x[\pi^2 - \nabla^2\phi + m^2\phi^2].$$

Conventionally, the system can be quantized by treating fields as operators and prescribing an appropriate algebra. This is done by choosing foliation of the space-time into space-like hypersurfaces keeping time fixed and using equal time commutation relations. It is interesting to note that, in general, in non-Minkowskian space-time this foliation possesses some difficulties which has also been circumvented in Schrödinger picture formalism.

3 Basic Formalism

In the Schrödinger picture approach, we take the basis vectors of the state vector space to be the eigenstates of the field operator $\phi(x)$ on a fixed time hypersurface with eigenvalue $\psi(\bar{x})$. The quantum mechanical state $|\psi(t)\rangle$ is replaced by a functional of the c-number field

$$|\psi(t)\rangle \longrightarrow \Psi[\phi, t].$$

Inner products are represented by functional integration and operators are represented by functional kernel. The action of an operator can be realized as a product and that of a canonical momentum as a functional differentiation. Thus, we have the following fundamental relations:

$$\langle\psi_1|\psi_2\rangle = \int \mathcal{D}\phi\, \psi_1^*(\phi)\psi_2(\phi)$$

$$\mathcal{O}\,|\psi\rangle = \int \mathcal{D}\phi\, \mathcal{O}(\phi, \tilde{\phi})\psi(\phi))$$

$$\Phi(x)\,|\psi(t)\rangle \longrightarrow \Psi(x)\Psi[\phi, t]$$

$$\Pi(x)\,|\psi(t)\rangle \longrightarrow -i\frac{\delta}{\delta\phi(x)}\Psi[\phi, t].$$

The dynamical evolution of a given initial state between the space-like hypersurfaces is described by the functional Schrödinger equation. This equation can be obtained by defining a time-dependent effective action and impose the condition that $|\Psi(t)\rangle$ is stationary against arbitrary variations.

$$\Gamma = \int dt\, \langle\Psi(t)|i\partial_t - H|\Psi(t)\rangle \tag{1}$$

$$i\frac{\partial\Psi(\phi,t)}{\partial t} = H\Psi(\phi,t) = \int_x \left[-\frac{1}{2}\frac{\partial^2}{\partial\phi^2(x)} + \frac{1}{2}(\nabla\phi)^2 + V(\phi)\right]\Psi(\phi,t). \quad (2)$$

For a time-dependent Hamiltonian, a separation of variables gives

$$\Psi_t[\phi] = e^{-iEt}\Psi[\phi]$$

and we are left with the functional eigenvalue problem

$$H\left(\phi, -i\frac{\partial}{\partial\phi}\right)\Psi[\phi] = E\Psi[\phi].$$

4 Example of Non-interacting Scalar Field

We have $V(\phi) = \frac{1}{2}m^2\phi^2$ and the free field Hamiltonian can be written as

$$H = \frac{1}{2}\int_x \left(\frac{\delta^2}{\delta(\phi(x)\delta\phi(x)} + \phi(x)(-\nabla^2 + m^2)\phi(x)\right). \quad (3)$$

We choose a Gaussian functional as the ground state wave functional, and by substituting it in the Schrödinger equation and performing the second-order functional derivative, we get specific expressions for the width of the Gaussian (G) and ground state energy eigenvalue.

$$\Psi_0[\phi] = \mathcal{N}exp - \frac{1}{2}\left(\int_{x,y}\phi(x)G(x,y)\phi(y)\right) \quad (4)$$

$$\left(\int_{x,y}\frac{1}{2}G(x,x)\delta(x-y) - \phi(x)G^2(x,y)\phi(y) + \phi(x)(-\nabla^2 + m^2)\phi(x)\right)$$
$$\times\ \Psi_0(x) = E_0\Psi_0[\phi] \quad (5)$$

$$G^2(x,y) = (-\nabla^2 + m^2)\delta(x-y) \quad (6)$$

$$E_0 = \frac{1}{2}\int_x G(x,x). \quad (7)$$

The bi-local kernel G represent the width of the Gaussian functional and E_0 represent the zero-point energy or vacuum energy of the free scalar field. Above quantities can be written more conveniently in momentum space by taking the Fourier transform to get

$$G(x-y) = \frac{1}{2\pi^3} \int_p e^{ip(x-y)} \sqrt{p^2+m^2} \tag{8}$$

$$E_0 = \frac{1}{2\pi^3} \int_x \int_p \frac{1}{2} \sqrt{p^2+m^2} \tag{9}$$

$$\Psi_0[\hat{\phi}] = Cexp(\frac{1}{2\times 2\pi^3}) \int_p \hat{\phi}(p)\hat{\phi}(-p)\sqrt{p^2+m^2} \tag{10}$$

$$\sqrt{p^2+m^2} \equiv \omega_p. \tag{11}$$

By using the normalization condition we will get the normalization constant

$$C = \prod_p \left(\frac{\omega_p}{\pi}\right)^{\frac{1}{4}}.$$

Following the procedure usually used in conventional quantum mechanics, we can define the creation and annihilation operators to obtain the excitations of vacuum state. In momentum space, they are defined as

$$a(\hat{\mathbf{p}}) = \frac{1}{\sqrt{2}} \int_x e^{i\mathbf{p}\cdot(x-y)} \left(\omega^{\frac{1}{2}}(\mathbf{p})\hat{\phi}(\mathbf{x}) + i\omega^{\frac{1}{2}}(\mathbf{p})\hat{\pi}(\mathbf{x})\right) \tag{12}$$

$$a(\hat{\mathbf{p}}) = \frac{1}{\sqrt{2}} \int_x e^{-i\mathbf{p}\cdot(x-y)} \left(\omega^{\frac{1}{2}}(\mathbf{p})\hat{\phi}(\mathbf{x}) - i\omega^{\frac{1}{2}}(\mathbf{p})\hat{\pi}(\mathbf{x})\right). \tag{13}$$

These operators satisfy the standard commutation relations

$$\left[a(\mathbf{p}), a^{\dagger}(\mathbf{p}')\right] = (2\pi)^3\delta(\mathbf{p}-\mathbf{p}') \tag{14}$$

and the Hamiltonian takes the form

$$H = \int \frac{d^3\mathbf{p}}{(2\pi)^3}\omega(\mathbf{p})a^{\dagger}(\mathbf{p})a(\mathbf{p}) + \frac{1}{2}\int \frac{d^3\mathbf{p}}{(2\pi)^3}\omega(\mathbf{p})(2\pi)^3\delta^3(\mathbf{p}-\mathbf{p}'). \tag{15}$$

To obtain the excited states, we need to apply the creation operator to the ground state wave functional. This is realized in practice by representing the creation and annihilation operator in functional derivative representation.

$$a(\mathbf{p}) = \frac{1}{\sqrt{2}} \int_x e^{i\mathbf{p}\cdot(x-y)} \left(\omega^{\frac{1}{2}}(\mathbf{p})\phi(x) + i\omega^{-\frac{1}{2}}\frac{\delta}{\delta\phi(x)}\right) \tag{16}$$

$$a^{\dagger}(\mathbf{p}) = \frac{1}{\sqrt{2}} \int_x e^{-i\mathbf{p}\cdot(x-y)} \left(\omega^{\frac{1}{2}}(\mathbf{p})\phi(x) - i\omega^{-\frac{1}{2}}\frac{\delta}{\delta\phi(x)}\right) \tag{17}$$

$$a(\mathbf{p})\Psi_0[\phi] = 0.$$

First excited state

$$\Psi_1(\phi) = \mathcal{C}a^\dagger(\mathbf{p}_1)\Psi_0(\phi)$$

represents a state with one scalar particle of mass m and momentum p_1 and energy ω_{p_1}. Even though the procedure given above analogous to conventional quantum mechanics is exact for free fields the solution of Schrödinger equation in functional form for interacting fields requires approximations. Most commonly used approximation is the variational approximation which we apply for the interacting fields.

5 Example of Φ^4 Field

As the trial wave functional, we may use the same Gaussian wave functional discussed above which is parametrized by unknown parameters which will be varied and we obtain self-consistent equations for these parameters which can be solved. We choose a Gaussian trial state for implementing variational approximation which is in direct analogy with standard harmonic oscillator solutions in quantum mechanics. The most general form of the Gaussian wave functional including the variational parameter is given below:

$$\Psi_t[\phi] = exp\left(-\int_{xy}(\phi(x) - \hat{\phi}(x,t)\Omega(x,y,t)(\phi(y) - \hat{\phi}(y,t)\right.$$
$$\left.+i\Pi(x,t)(\phi(x) - \hat{\phi}(x,t)\right). \quad (18)$$

On separating the variational parameter Ω into real and imaginary parts

$$\Omega = \frac{G^-1(x,y,t)}{4} - i\Sigma(x,y,t),$$

we get

$$\Psi(\phi,t) = exp\left[-\int_{x,y}(\phi(x) - \hat{\phi}(x,t))\left[\frac{G^{-1}(x,y,t)}{4} - i\Sigma(x,y,t)\right]\right.$$
$$\left.\times(\phi(y) - \hat{\phi}(y,t)) + i\int_x \hat{\Pi}(x,t)[\phi(x) - \hat{\phi}(x,t)\right]. \quad (19)$$

In the above functional, the Gaussian is centred at $\hat{\phi}$ and the width is G. Σ and Π play the role of conjugate momentum for G and $\hat{\phi}$. We choose the variational parameters as $\hat{\phi}$, $\hat{\Pi}$, G and Σ are variational parameters as well as expectation values.

$$\langle\phi(x)\rangle = \hat{\phi}(x,t) \quad (20)$$

$$\left\langle -i\frac{\delta}{\delta\phi(x)}\right\rangle = \Pi(\hat{x},t) \tag{21}$$

$$\langle\phi(x)\phi(y)\rangle = \phi(\hat{x},t)\hat{y},t) + G(x,y,t) \tag{22}$$

$$\left\langle i\frac{\partial}{\partial t}\right\rangle = \int_x \Pi(\hat{x},t)\dot{\hat{\phi}}(x,t) + \int_{x,y}\Sigma(x,y,t)\dot{G}(x,y,t) \tag{23}$$

$$V^{(n)}(\phi) \equiv \frac{d^n V(\phi)}{d\hat{\phi}^n}.$$

To apply this formalism to the scalar field having quartic self-interaction, the expression for effective action can be written up to two-loop level. Obviously the effective action will consist of a classical action term and higher order loop corrections.

$$\begin{aligned}\Gamma = \int dt\Bigg[\int_x \hat{\Pi}\dot{\hat{\phi}} - \frac{1}{2}(\nabla\hat{\phi})^2 - V(\hat{\phi}) + \int_{x,y}\Sigma\dot{G} - 2\int_{x,y,z}\Sigma G\Sigma \\ - \int_x \frac{1}{8}G^{-1}(x,x,t) - \frac{1}{2}\nabla_x^2 G(x,y,t)|_{x=y} + \frac{1}{2}V^{(2)}(\hat{\phi})G(x,x,t)\Bigg] \\ - \frac{1}{8}V^{(4)}(\hat{\phi})\int_x G(x,x,t)^2.\end{aligned} \tag{24}$$

Variations of the parameters are performed through functional derivatives to obtain the time derivatives of the variational parameters describing the Gaussian trial function. This is a functional extension of what we do in elementary quantum mechanics. Identifying the first term as the classical action and performing variations, we get

$$\frac{\delta\Gamma}{\delta\dot{\hat{\phi}}(x,t)} = 0 \to \dot{\hat{\Pi}}(x,t) = \nabla_x^2\hat{\phi}(x,t) - V^{(1)}(\hat{\phi}) - \frac{1}{2}V^{(3)}(\hat{\phi})G(x,x,t) \tag{25}$$

$$\begin{aligned}\frac{\delta\Gamma}{\delta\dot{\hat{\Pi}}(x,t)} &= 0 \to \dot{\Sigma}(x,y,t) + 2\int_x \Sigma(x,z,t)\Sigma(x,y,t) \\ &= \frac{1}{8}G^{-2}(x,y,t) + \left[\frac{1}{2}\nabla_x^2 - \frac{1}{2}V(2)(\hat{\phi}) - \frac{1}{4}V^{(4)}(\hat{\phi})G(x,x,t)\right]\delta^{\nu}(x-y)\end{aligned} \tag{26}$$

$$\frac{\delta\Gamma}{\delta\Sigma(x,y,t)} = 0 \to \dot{G}(x,y,t) = 2\left[\int_x G(x,z,t)\Sigma(x,y,t) + \Sigma(x,z,t)G(z,y,t)\right]. \tag{27}$$

Using the above equations we can approximately determine the vacuum state of ϕ^4 theory. We take $\hat{\phi}$ to be x independent. Kernels can be expressed in momentum space using Fourier transform

$$G(x, y) = \int \frac{d^3\mathbf{p}}{(2\pi)^3} e^{i\mathbf{p}\cdot(x-y)} \tilde{G}(\mathbf{p}).$$

For the vacuum functional in flat space, time Kernels are time independent which implies that $\dot{\pi} = \Sigma = 0$. The inverse Fourier transform of the Kernel is

$$\tilde{G}(\mathbf{p}) = \frac{1}{2\sqrt{P^2 + M^2}}$$

with

$$M^2 = m^2 + \frac{\lambda}{2}\phi^2 + \frac{\lambda}{2}\int \frac{d^3\mathbf{p}}{(2\pi)^3}\tilde{G}(\mathbf{p})$$

when $\lambda = 0$ this reduces to the free field theory. M is usually called the effective mass.

The effective potential in this model is obtained as

$$V_{eff}(\tilde{\phi}) = \frac{1}{2}m^2\phi^2 + \frac{\lambda}{4!}\tilde{\phi}^4 + \frac{1}{4}G^{-1}(x, x) + \frac{\lambda}{8}G(x, x)G(x, x). \tag{28}$$

G satisfies the equation

$$\frac{1}{4}G^{-2}(x, y) = -\nabla^2 + m^2 + \frac{\lambda}{2}\tilde{\phi}^2 + \frac{\lambda}{2}G(x, x)\delta^3(x - y).$$

Defining Λ as a mass scale the effective potential can be written as

$$\begin{aligned} V_{eff}(\tilde{\phi}) &= \frac{1}{2}m^2\tilde{\phi}^2 + \frac{\lambda}{4!}\tilde{\phi}^4 - \frac{1}{2\lambda}\left(M^2 - m^2 - \frac{\lambda}{2}\tilde{\phi}^2\right)^2 \\ &\quad + \frac{M^2}{2}I_1 - \frac{M^4}{2}I_2(\Lambda) + \frac{M^4}{64\pi^2} \end{aligned}$$

$$I_1 \equiv \int \frac{d^3p}{(\pi)^3}\frac{1}{2|p|}$$

$$I_(\Lambda) \equiv \int \frac{d^3p}{(\pi)^3}\left(\frac{1}{2|p|} - \frac{1}{2\sqrt{p^2 + \Lambda^2}}\right)$$

$$M^2 = m + \frac{\lambda}{2}\tilde{\phi}^2 - \frac{\lambda}{2}M^2 I_2 + \frac{\lambda}{32\pi^2}M^2 ln\frac{M^2}{\Lambda^2}.$$

We need to renormalize the divergent quantities I_1 and I_2. This is achieved by redefining the independent parameters m and λ which becomes m_R and λ_R. The effective potential becomes finite if we use the following renormalization prescription:

$$\frac{dV_{eff}}{dM^2} = \frac{m_R^2}{\lambda_R}$$

$$\frac{d^2V_{eff}}{d(M^2)^2} = \frac{1}{3\lambda_R}$$

which can be written as

$$\frac{m_R^2}{\lambda_R} = \frac{m^2}{\lambda} + \frac{1}{2}I_2$$

$$\frac{1}{\lambda_R} = \frac{1}{\lambda} + \frac{1}{2}I_2$$

Giving the effective potential and effective mass which are finite.

$$V_{eff} = -\frac{M^4}{2\lambda_R} + \frac{M^2}{2}\tilde{\phi}^2 + \frac{m_R^2}{\lambda_R}M^2$$

$$+\frac{M^4}{64\pi^2}\left[ln\frac{M^2}{\Lambda^2} - \frac{1}{2}\right]$$

$$M^2 = m_R^2 + \frac{\lambda_R}{2}\tilde{\phi}^2 + \frac{\lambda_R}{3\pi^2}M^2 ln\frac{M^2}{\Lambda^2}$$

It is interesting to note that in quantum mechanics the variational approximation using Gaussian state produces very accurate results. But in field theory the results are not extremely accurate. Using other wave functionals like coherent state wave functional may produce more accurate results which is to be verified.

6 Effective Action for ϕ^6 Model

The procedure developed above for ϕ^4 model can be extended to ϕ^6 model. The coupling effects will show up only at three-loop levels and we write the expansions up to three loops.

$$\Gamma = \int dt \Bigg[\int_x \hat{\Pi}\dot{\hat{\phi}} - \frac{1}{2}(\nabla\hat{\phi})^2 - V(\hat{\phi}) + \int_{x,y} \Sigma\dot{G} - 2\int_{x,y,z} \Sigma G \Sigma$$
$$- \int_x \frac{1}{8}G^{-1}(x,x,t) - \frac{1}{2}\nabla_x^2 G(x,y,t)|_{x=y} + \frac{1}{2}V^{(2)}(\hat{\phi})G(x,x,t)$$
$$-\frac{1}{8}V^{(4)}(\hat{\phi})\int_x G(x,x,t)^2 - \frac{1}{16}V^{(6)}(\hat{\phi})\int_x G^3(x,x,t)\Bigg]. \tag{29}$$

Following the procedure used in section 5 we identify the first term in equation (25) as classical action and perform variations with the variational parameters to obtain

$$\frac{\delta\Gamma}{\delta\dot{\hat{\phi}}(x,t)} = 0 \rightarrow \dot{\hat{\Pi}}(x,t) = \nabla_x^2\hat{\phi}(x,t) - V^{(1)}(\hat{\phi}) - \frac{1}{2}V^{(3)}(\hat{\phi})G(x,x,t) - \frac{1}{8}V^{(5)}(\hat{\phi})G^2(x,x,t) \tag{30}$$

$$\frac{\delta\Gamma}{\delta\dot{\hat{\Pi}}(x,t)} = 0 \rightarrow \dot{\Sigma}(x,y,t) + \int_x \Sigma(x,z,t)\Sigma(x,y.t) = \frac{1}{8}G^{-2}(x,y,t) + \left[\frac{1}{2}V^{(2)}(\hat{\phi}) - \frac{1}{4}V^{(4)}(\hat{\phi})G(x,x,t) - \frac{1}{4}V^{(6)}(\hat{\phi})G(x,x,t)\right]\delta^\nu(x-y) \tag{31}$$

$$\frac{\delta\Gamma}{\delta\Sigma}(x,y,t) = 0 \rightarrow \dot{G}(x,y,t) = 2\left[\int_x G(x,z,t)\Sigma(x,y,t) + \Sigma(x,z,t)G(z,y,t)\right]. \tag{32}$$

7 Static Effective Potential

Since the renormalization of time-dependent effective potential proceeds along the same line as that of static effective potential, we evaluate the static effective potential for ϕ^6 model. It can be written in the form

$$V_{eff}(\hat{\phi},G) = \frac{1}{2}m^2\hat{\phi}^2 + \frac{\lambda}{4!}\hat{\phi}^4 + \frac{\xi}{6!}\hat{\phi}^6 + \left(\frac{1}{2}m^2 + \frac{\lambda}{4}\hat{\phi}^2 + \frac{\xi}{48}\hat{\phi}^4\right)G(x,x) + \left(\frac{\lambda}{8} + \frac{\xi}{16}\hat{\phi}^2\right)G(x,x) + \frac{\xi}{48}G^3(x,x) + \frac{1}{8}tr\,G^{-1}(x,x) - \frac{1}{2}\nabla_x^2 G(x,x). \tag{33}$$

By performing the variation with respect to G the gap equation will be obtained as

$$\frac{1}{4}G^{(-2)}(x,y) = \left[-\nabla_x^2 + m^2 + \frac{\lambda}{2}\hat{\phi}^2 + \frac{\xi}{24}\hat{\phi}^4 + \frac{\lambda}{2}G(x,y) + \frac{\xi}{4}\hat{\phi}^2G^2(x,y) + \frac{\xi}{48}G^3(x,y)\right]\delta(x-y). \tag{34}$$

We assume translational invariance and define the Fourier transform

$$\int \frac{d^\nu k}{2\pi^\nu} e^{i\mathbf{k}\cdot x} f(\mathbf{k})$$

$$G(x,x)=\int\frac{d^{\nu}}{2\pi^{\nu}}\frac{1}{2}\left[\left[k^2+m^2+\frac{\lambda}{2}\hat{\phi}^2+\frac{\xi}{24}\hat{\phi}^4\right]\right.$$
$$\left.\times\frac{\lambda}{2}G(x,x)+\frac{\xi}{4}\hat{\phi}^2G(x,x)+\frac{\xi}{48}G(x,x)\right]^{-\frac{1}{2}} \tag{35}$$

$$G(x,x)=\int\frac{d^{\nu}k}{2\pi^{\nu}}\frac{1}{2(k^2+M^2)^{\frac{1}{2}}}.$$

In the above equation, an Anzatz is fixed for G in terms of an effective mass M which can be treated as a variational parameter which is $\hat{\phi}$ dependent. The static effective potential then can be written as

$$V_{eff}(\phi,M)=\frac{1}{2}\int\frac{d^{\nu}k}{2\pi^{\nu}}\sqrt{(K+M^2)}+\left(\frac{1}{2}m^2\hat{\phi}_2+\frac{\lambda}{4!}\hat{\phi}^4+\frac{\xi}{6!}\phi^6\right)$$
$$+\frac{1}{2}\left[M^2-m^2-\frac{\lambda}{2}\hat{\phi}^2-\frac{\xi}{24}\hat{\phi}^4\right]G(x,x)$$
$$+\left[\frac{\lambda}{8}+\frac{\xi}{16}\hat{\phi}^2\right]+\frac{\xi}{48}G^3(x,x). \tag{36}$$

The equation (36) shows that the effective potential expression obtained here is the same as the one obtained using Gaussian effective potential approach. This is only natural since when time dependence of effective action is not taken into account definitions of effective action in both approaches coincide. It also becomes clear that the formalism is equivalent to CJT approach at zero temperature apart from a term $\xi\phi^2$ which does not contribute when daisy and super daisy graphs alone are taken into account. At ϕ^4 level both approaches coincide.

Considering the first and second terms alone of Eq. (36) it can be seen that one loop effective potential result is contained in the expression with the mass term replaced by the effective mass. Identity with Gaussian effective potential becomes more transparent if we make the following correspondence in notation.

$$G(x,x)\rightarrow I_0=\int\frac{d^{\nu}k}{2\pi^{\nu}}\frac{1}{2(\sqrt{(k^2+m^2}}$$

$$\frac{1}{2}\int\frac{d^{\nu}k}{2\pi^{\nu}}\sqrt{k^2+m^2}\rightarrow I_1$$

$$M\rightarrow\Omega.$$

Since the effective potential is an ordinary function (not a functional), stationary requirements with respect to ϕ and M^2 is obtained by ordinary differentiation.

$$\frac{\partial V}{\partial \phi} = \phi\left[m^2 + \frac{\lambda}{6}\phi^2 + \frac{\xi}{120}\phi^4 + \frac{\lambda}{2}G(x,x) + \frac{\xi}{12}\phi^2 G(x,x)\right.$$
$$\left.+\frac{\xi}{4}\phi^2 G(x,x)\right] = 0 \tag{37}$$

$$\frac{\partial V}{\partial M^2} = -\frac{1}{2}\left[M - m^2 - \frac{\lambda}{2}\phi^2 - \frac{\xi}{24}\phi^4 - \frac{\lambda}{4}\phi^2 G(x,x)\right.$$
$$\left.-\frac{\xi}{8}G(x,x)G(x,x)\right]\frac{\partial G(x,x)}{\partial M^2} = 0. \tag{38}$$

Conventional effective potential is defined as the solution of Eq. (38). The effective mass is given by

$$M^2(\phi) = \left[m^2 + \frac{\lambda}{2}\phi^2 + \frac{\xi}{24}\phi^4 + \frac{\lambda}{2}G(x,x) + \frac{\xi}{8}G(x,x)G(x,x)\right]. \tag{39}$$

Required expression for the effective mass is obtained simply by replacing M by $M(\phi)$ in Eq. (36). This equation exhibits certain very important features of ϕ^4 and ϕ^6 theories relevant at zero temperature.

$$V'(\phi) = \phi\left[M^2(\phi) - \frac{\lambda}{3}\phi^2\right]. \tag{40}$$

For ϕ^4 theory, $M^2(\phi)$ is intrinsically positive. Hence, for negative coupling constant, the only solution to the above equation is $\phi = 0$ which means that the potential is unbounded from below. In the case of ϕ^6 theory

$$V'(\phi) = \phi\left[M^2(\phi) - \frac{\lambda}{3}\phi^2 + \frac{\xi}{30}\phi^4\right] - \frac{\xi}{4}\left[1 - \frac{\phi^2}{3}\right]G^2. \tag{41}$$

Non-zero turning points are possible also for negative coupling constant λ. The ϕ^6 theory in Hartree–Fock approximation require up to three loops for exhibiting the effects of ξ coupling. Thus, we have four parts for the effective potential

$$V_{eff}(\phi, M(\phi)) = V^0 + V^1 + V^2 + V^3 \tag{42}$$

$$V^0 = \left[\frac{1}{2}m^2\phi^2 + \frac{\lambda}{4!}\phi^4 + \frac{\xi}{6!}\phi^6\right] \tag{43}$$

$$V^1 = \frac{1}{2}\int \frac{d^3k}{(2\pi)^3} ln[k^2 + M^2(\phi)] \tag{44}$$

$$V^2 = -\frac{\lambda}{8}G(x,x)G(x,x) - \frac{\xi}{16}(\phi)^2 G(x,x)G(x,x) \tag{45}$$

$$V^3 = -\frac{\xi}{24} G(x,x)G(x,x)G(x,x). \tag{46}$$

Above set of equations can be used to obtain the effective potentials for both ϕ^4 and ϕ^6 theories. In the following, we ignore the ϕ^2 dependence in order to establish equivalence with Hartree–Fock approximation usually used in CJT formalism.

8 Renormalization

Following the renormalization prescription developed for ϕ^4 theory, we can renormalize the ϕ^6 theory also. The effective mass $M(\phi)$ defining the effective potential is divergent due to the divergence of the kernel G(x,x). We define

$$G(M(\phi)) \equiv -\frac{M(\phi)}{4\pi}$$

$$E \equiv \sqrt{k^2 + M^2(\phi)}.$$

In (2 + 1) dimension coupling constant renormalization is not required. The renormalized mass is defined as

$$m_R^2 \equiv m^2 + \frac{1}{2}\lambda I_1 + \frac{\xi G(M(\phi))}{4} I_1 + \frac{\xi}{8} I_1^2$$

$$I_1 \equiv \int \frac{d^2k}{2\pi^2}\frac{1}{2k} = \lim_{\Lambda\to\infty}(\frac{\Lambda}{4\pi})$$

$$G(x,x) = \int \frac{d^2k}{2\pi^2}\frac{1}{2E}.$$

By actual evaluation introducing a cutoff parameter Λ we get

$$G(x,x) = G(M(\phi)) + I_1$$

which shows that $G(M(\phi))$ is the finite part of the vacuum propagator. A finite expression for the effective mass is obtained by expressing it in terms of the renormalized parameters.

$$M^2(\phi) = -m^2 + \frac{\lambda}{2}\phi^2 + \frac{\xi}{24}\phi^4 + \frac{\lambda}{2}G(M(\phi)) + \frac{\xi}{8}G(M(\phi))G(M(\phi)). \tag{47}$$

Second derivative of the tree-level potential is defined as m_{tree}

$$M^2(\phi) = m_{tree}^2 + \frac{\lambda}{2}G(M(\phi)) + \frac{\xi}{8}G(M(\phi))G(M(\phi)) \tag{48}$$

$$V^1(M(\phi)) = -\frac{M^3}{6\pi} + \frac{\Lambda^3}{6\pi}. \tag{49}$$

It is observed that V^1 depends on the cutoff parameter. Cancellation of this divergence is obtained by combining V^0, V^2 and V^3

$$V(M) = -\frac{M^3}{6\pi} + \frac{M^4}{2\lambda} - \frac{1}{2}M^2G(M) - \frac{M^2}{24\eta}G^2(M) - F(\phi) \tag{50}$$

with $\eta^{-1} = \frac{\xi}{\lambda}$ and

$$F(\phi) = \left[\frac{m^2}{4\eta} - \frac{\lambda}{12} - \frac{1}{6!}\xi\phi^2\right]\phi^4$$

$\xi = 0$ reproduces the ϕ^4 potential reported by Camellia and Pi at zero temperature. Using the unrenormalized gap equation we combine V^0, V^1, V^2 and V^3 and writing them in terms of renormalized parameters, we get

$$\begin{aligned} V^0 + V^1 + V^2 + V^3 = \frac{\lambda_R}{8}\left[\phi - \frac{2m_R^2}{\lambda}\right]^2 - \frac{\eta}{24}m_R^2\phi^4 \\ - \frac{\lambda_R}{8}G^2(M) - \frac{\xi_R}{48G^3(M) - F(\phi)} \end{aligned} \tag{51}$$

In the case of $(2+1)$-dimensional ϕ^4 theory $F(\phi) = \frac{\lambda}{12}$ which is finite. Thus, unlike (3 + 1)-dimensional ϕ^4 theory, the effective potential does not contain any unrenormalized parameters. But in the case of ϕ^6 theory $F(\phi)$ contain m which is an unrenormalized parameter. But here we can make $F(\phi) = 0$ by adjusting the parameters suitably and make the unrenormalized parameter vanish.

9 Conclusions

The static effective potential is obtained from the effective action for free fields and ϕ^4 and ϕ^6 scalar field theories using functional Schrödinger picture formalism with Gaussian functional variational approximation. It is shown that for (2 + 1)-dimensional ϕ^6 theory turning points can exist for both positive and negative coupling constant λ but in ϕ^4 theory turning points exist only for positive λ. This implies that ϕ^4 theory is unbounded from below.

Appendix

Most of the equations developed in the present review rely mainly on the concept of functional derivative which is briefly described in this appendix for the use of non-specialist reader. The functional $F[\phi]$ is a mapping from a normed linear space of functions 9 a Banach space to the field of real or complex numbers. As in ordinary calculus, functional derivative can also be expressed as limit of divided differences. The increment of function $\phi(x)$ localized at some arbitrary point y is

$$\delta\phi(x) = \eta\delta(x - y)$$

where $\delta(x - y)$ is the Dirac delta function. For practical purposes, we use the following definition of functional integral

$$\frac{\delta F[\phi]}{\delta\phi(y)} = \lim_{\eta \to 0} \frac{F[\phi + \eta\delta(x - y) - F[\phi]]}{\eta}.$$

Product rule and chain rule can be extended to functional derivatives also. If $F[\phi] = g[\phi] + H[\phi]$, we have

$$\frac{\delta F[\phi]}{\delta\phi(x)} = \frac{\delta G[\phi]}{\delta\phi(x)} H[\phi] + g[\phi] \frac{\delta F[\phi]}{\delta\phi(x)}$$

$$\frac{\delta}{\delta\phi(x)} F[G[\phi]] = \int dx \frac{\delta F[G[\phi]]}{\delta G(x)} \frac{\delta G[\phi]}{\delta\phi(y)}$$

and a simple example is considered here to make the procedure clear. The same procedure can be used to obtain functional derivatives of wave functional involving kernels discussed in this review.

$$F[\phi] = \int dx(\phi(x))^n$$

$$\frac{\delta F[\phi]}{\delta\phi(y)} = \lim_{\eta \to 0} \frac{\int dx(\phi(x) + \eta\delta(x - y)) - \int dx(\phi(x))^n}{\eta}$$
$$= \int dx\eta n\phi(x)^{n-1}\delta(x - y) = n(\phi(y))^{n-1}.$$

References

1. K.P. Satheesh, K. Babu Joseph, Pramana **49**(6), 591–601 (1997)
2. D.V. Long, G.H. Shore, Nucl. Phy. B **530**, 247, arXiv:hep-th/9065004 (1998)
3. H.C. Reis, Int. J. Mod. Phys. **A14**, 6029, arXiv:hep-th/0108175 (1999)
4. D.V. Long, G.M. Shore, Phy. B **530**, 279–303, arXiv:gr-qc/9607032 (1998)

5. K.P. Satheesh, K. Babu Joseph, Pramana **50**(2), 133–148 (1998)
6. J. Liu, C.-M. Sou, Y. Wang, arXiv:hep-th/1608.07909 (2016)
7. E. Nelson, arXiv:gr-qc/1601.03734 (2016)
8. S. Krug, arXiv:hep-th/1404.7005 (2014)
9. I.V. Kanachikov, *AIP Conference Proceedings 1514*, arXiv:gr-qc:/1212-6963 (2012)
10. A.H. Guth, S.Y. Pi, Phys. Rev. D **32**, 1899 (1985)
11. J. Kulson, Quant. Grav. **20**, 4285, arXiv:hep-th/0307079 (2003)
12. C.K. Sim, S.K. You, arXiv:cond-mat/012557
13. D.K. Kim, C.K. Kim, J. Phys. A **31**, 6029–6036, arXiv:hep-th/970992 (1998)
14. S.J. Hyun, G.H. Lee, J.H. Yee, Phys. Rev. D **50**, 6542, arXiv:hep-th/9406070 (1994)
15. B. Batfield, *Addison-Wesley* (1992)
16. R. Jackiw, *World Scientific* (1995)
17. T. Vachaspati, Quant. Grav. **26**, 215007, arXiv:gr-qc/0711.0006 (2019)
18. D.V. Long, G.M. Shore, arXiv:hep-th/9605004 (1996)

On Gravity: Remembrance of the Association Between a Guru and His Student/Colleague

Titus K. Mathew

Abstract In 1988, Prof. K Babu Jospeh and Prof. M Sabir have reformulated the Einstein's gravity as a flat space gauge theory of $U(2) \times U(2)$ symmetry. This becomes an important step towards the dream of creating the quantum version of gravity. We are presenting a brief review of their paper.

Keywords Gravity · Space–time · Euclidean geometry · Riemannian geometry

1 Introduction

One of the major areas of research in which Prof. Babu Joseph and his students, especially with Prof. M Sabir, pursued is gravity. Prof. Sabir was one among the first batch students of Prof. Babu Joseph. After taking Ph.D., Prof. Sabir joined as a faculty in the Department of Physics. Together they started working on gravity some time in 1984. The major problem that prevailed then, now also, is the merging of the classical gravity with quantum mechanics, in short quantum gravity theory. The classical theory by Einstein is a highly non-linear one, and moreover the fundamental variable is the metric or say the geometry of space–time itself, which rendered the major difficulty in achieving a well-defined quantum version of the classical theory of gravity. In this article, I plan to make a brief review, first about the Einstein theory of gravity and then about the magnificent paper published by these eminent people on the reformulation of gravity using gauge field theory, which becomes a classic approach that paved one of the ways towards a quantum gravity theory.

T. K. Mathew (✉)
Department of Physics, Cochin University of Science and Technology, Kochi, Kerala, India
e-mail: tituskmathew@gmail.com

K. S. Sreelatha and V. Jacob (eds.), *Modern Perspectives in Theoretical Physics*,
https://doi.org/10.1007/978-981-15-9313-0_5

2 A Very Brief History of Association Between Prof. Babu Joseph and Prof. Sabir

Prof. Sabir joined with Prof. Babu Joseph for Ph.D. in the year 1974. They together started working on field theory, a pioneer area in theoretical physics at that time. After Ph.D., Prof. Sabir joined the Department of Physics as a faculty in 1979 where his guru, Prof. Babu Joseph was a senior professor. This became the second spell of the continuation of their long-term association, which culminated into the fruitful research in many fields like field theory, non-linear dynamics, gravity, etc. Prof. Babu Joseph (KBJ) later became the Head of the same Department and then became the Vice-Chancellor of CUSAT. Prof. Sabir (MS) retired from service in the year 2011 and then continued as an Emeritus Professor there until his death on October 2019. Their association resulted into some remarkable publications in the field of field theory, non-linear dynamics and gravity. Among these, we will concentrate on gravity which acquired much appreciation from Theoretical Physicist community during that time.

3 On Gravity

As noted earlier the important contribution of KBJ and MS in gravity is their pioneer work on the reformulation of Einstein's gravity as flat space gauge theory. This work has been published as a paper in the reputed journal, Modern Physics Letters A in the year 1988. It was an important contribution towards the dream of creating a quantum theory of gravity.

Einstein's theory of gravity is a revolution in understanding gravity. Even though Newton's theory of gravity was very successful in predicting the Kepler's planetary laws, it fails in explaining the perihelion shift of planet Mercury accurately and also in understanding the effect of gravity on light. Around 1915, Einstein reformulated gravity in a completely new framework, in which the gravity is explained as a geometrical theory of space–time. The all time new idea was that matter can influence the geometry of the space–time around it and gravity is nothing but this effect of matter on space–time. Hence a correct equation of gravity is a relation between the modified geometry of the space–time and the energy/matter present in the given location. In the absence of gravity, the geometry of the space–time is similar to Euclidean and is often called as flat geometry. But, in the presence of matter, the geometry of the surrounding space–time will become non-Euclidian or say Riemannian and is referred to as curved geometry. The curvature of the space–time is represented by Einstein tensor, G_{ab} which is then equated to the matter and energy represented by a second rank tensor, T_{ab}. The new law of gravity is then stated as

$$G_{ab} = \kappa T_{ab}, \tag{1}$$

where κ depends on G the fundamental constant of gravity, first used by Newton. Even though this is a simple looking equation, in the inside it is highly complicated. The exact form of the Einstein tensor is $G_{ab} = R_{ab} - (1/2)g_{ab}R$, where R_{ab} is a second rank tensor called Ricci tensor which represents the curvature of space–time and R is the scalar corresponding to it. The object g_{ab} is known as the metric components of the space–time which is the basic quantity carrying the geometry of the given space–time. The immediate question that may arise is "if R_{ab} is representing the curvature of the space–time, then what is the reason for using another second rank tensor G_{ab} for representing the same curvature of the space-time in writing down the Einstein equation?". The reason is the following. If one uses R_{ab} alone in equation (1) instead of G_{ab}, then the equation will posses the problem of unequal divergence on both sides of the equation. That is the divergence of R_{ab} is not equal to zero while divergence of T_{ab} is zero. It was Einstein's intelligence which ultimately frame the left-hand side of the equation with G_{ab}, a second rank tensor, representing curvature and at the same time posses zero divergence. In the absence of matter, the equation of gravity will reduces to $R_{ab} = 0$ under linear limit. The immediate success of Einstein's equation is that it explains the perihelion shift of planet Mercury accurately and predicts the effect of gravity on light that the light will bend under gravity. The bending of light was later observationally verified by a historic event by a team led by Eddington.

Even though the success of reformulating gravity as the curvature of space–time is remarkable, mathematically the situation becomes more complicated than that of Newton's law. The difficulty is that the quantity R_{ab} contains second-order partial derivatives of g_{ab} and also the products of metric components, which makes the equation highly non-linear. So finding solutions to Einstein's equation becomes a Himalayan task. Schwarzschild comes with the first solution for compact mass distribution, where he took advantage of the simple spherically symmetric distribution of matter, which simplified the process of deriving the solution a lot. The other prominent situation in which the equation possess a well-defined solution is for an isotropic and homogeneous distribution of matter, and is the case for our universe, which was first obtained by Friedmann and later by Lamaitre, Robertson and walker. Unless there exists convenient symmetries, it becomes quite impossible to find an analytical solution for Einstein's field equation.

3.1 Searching for Quantum Version of Gravity

Gravity is a peculiar force compared to other fundamental forces in the sense that it is the only one which can affect the geometry of the space–time. In essence a theory of gravity is a theory of the space–time geometry itself. In the later part of the twentieth century it was showed that all systems are basically quantum mechanical in nature. Hence theories of all fundamental laws of nature must be described in terms of the principles of quantum theory. Towards the end of the last century, the classical laws of electrodynamics got modified in to a description based on quantum principles, and thus we have quantum electrodynamics. Therefore, one expects that gravity should also be modified to fit into the framework of quantum theory and such a unification is what we called as the quantum gravity. The weak and strong forces were reformulated according to the quantum principles. Because of this, electromagnetic and weak nuclear interactions are unified within the Standard Model and strong nuclear interaction is currently described by Quantum Chromodynamics. But, classical General Relativity by Einstein steadily escaped the attempts of quantisation. It is well known that the problem arises because of the non-renormalizability of General Relativity. It was shown that General Relativity theory is non-renormalizable after the inclusion of matter fields. The difficulty with non-renormalizable theories is that they are not predictive, since to make well-defined predictions it potentially requires an infinite number of divergent renormalizations.

Another problem is associated with the question of unitarity. According to quantum mechanics, knowledge about the state of a system at one instant is equivalent to the knowledge about its state at any other instant. This is due to a one-to-one correspondence, between the states at two different instants, induced by the evolution equations, the Schrodinger equation which is a linear equation. But, the governing equation in classical gravity is the non-linear field equation, which indicates that the evolution of a state in gravity may not be unitary. For example, consider a black hole formed due to the collapse of a star, the resulting configuration is assumed to be a pure state. But it radiates out completely, leaving behind only the Hawking radiation, which is not a pure state. Such an evolution of a pure state into a mixed state is not allowed in quantum mechanics. Like this, there arose many barricades like, positivity, causality, etc., which I am not describing here, for that one may refer to some standard review article.

3.2 Reformulation of Gravity in Flat Space Using Gauge Field Theory—A Remarkable Contribution by KBJ and MS

It is known that the major issue in formulating the quantum version of gravity is its non-renormalizability. But it was noticed that at least in the currently attainable energies, gauge field theories are renormalizable. Various attempts have been done for formulating a gauge field version of Einstein's gravity. The major huddle is to find a gauge theoretical formulation of the space–time metric, especially the curvature, since gauge field theory was formulated in flat space. Attempts were made to gauge the Lorentz group in a local surrounding and then to transport over to the curved space manifold to form a gauge theory of gravitation, but it becomes fruitless. A possible way pointed out is to seek a gauge theoretical reformulation of gravity using Yang-Mills (YM) [1] type of action, because attempt using non-YM action leads to the undesirable features of non-metricity and torsion.

A major step in reformulating gravity as a gauge theory was done by Dehnen and Ghaboussi(DG) [2] in 1986, in which the authors created a gauge reformulation of the

Einstein gravity using Yang-Mills action based on $SU(2) \times U(1)$ group. However, this theory has some severe drawbacks. The major drawback is that the metric of the curved space is a composite one obtained from the basic YM fields and hence is not fundamental. Secondly, many of the assumptions made in their paper are completely arbitrary, hence of doubtful validity. In 1988, KBJ and MS [3] came up with an alternative theory, in which they have rectified the above two shortcomings of the DG theory, and have specifically showed that it is possible to reformulate Einstein's gravity using non-YM gauge fields in flat space based on $U(2) \times U(2)$ symmetry and it reduces to the usual YM form in the linear limit.

We will now try to follow the main arguments in the mentioned paper by KBJ-MS [3]. Consider a Minkowski space gauge theory based on the symmetry group $U(2) \times U(2)$ (this has been chosen arbitrarily). Let $A_i^{(a)}$, $B_i^{(a)}$ are a pair of gauge potentials compatible with the symmetry group $U(2)$, that is, each will transform according to the $U(2)$ group. The resulting field strength tensors can then be defined as

$$F_{ij}^a = \partial_i A_j^a - \partial_j A_i^a + e\epsilon^{abc} A_i^b A_j^c \tag{2}$$

$$G_{ij}^a = \partial_i B_j^a - \partial_j B_i^a + f\epsilon^{abc} B_i^b B_j^c \tag{3}$$

where e and f are gauge coupling constants and ϵ^{abc} are the structure constants of the symmetry group $U(2)$. The indices $a, b = 1, 2, 3, 4$ are the internal indices and i, j are Minkowski space indices. Following from here a fourth rank tensor can be constructed (following the method of tensor decomposition) as

$$\tilde{R}_{ijkl} = \left[K_{ij}^a H_{kl}^a + \frac{1}{2} K_{ik}^a H_{jl}^a + \frac{1}{2} K_{il}^a H_{kj}^a + K \leftrightarrow H \right]. \tag{4}$$

where $K_{ij}^a = (1/6)(F_{ij}^a + iG_{ij}^a)$ and $H_{ij}^a = -(1/6)(F_{ij}^a - G_{ij}^a)$ are two antisymmetric tensors. By nature this tensor satisfies the symmetries, $\tilde{R}_{ijkl} = -\tilde{R}_{jikl} = -\tilde{R}_{ijlk}$; $\tilde{R}_{ijkl} = \tilde{R}_{klij}$; $\tilde{R}_{ijkl} + \tilde{R}_{iljk} + \tilde{R}_{iklj} = 0$, which are symmetries similar to the curvature tensor in Einstein's gravity theory, but beware in the present case the tensor was formulated in flat space.

After constructing the tensor $\tilde{R}_{ijkl}$, the next argument posed by KBJ-MS is the most interesting one. They argued that the $U(2) \times U(2)$ gauge theory formulated above becomes the theory of gravitation. This argument of course needs further explanation. The authors substantiate their claim in an indirect way. It is well known that the classical Einstein's theory of gravitation is formulated in terms of the metric field of the curved space. So the attempt here is to reformulate the flat space $U(2) \times U(2)$ gauge theory in terms of the metric field of the curved space. For this, the authors put forwarded a clever postulate that the flat space gauge potentials exist in the flat neighbourhood of every point in the curved space. So effectively the metric properties of the curved space can be consistently described by the flat space gauge potentials. In this way, the flat space gauge potential is effectively linked with the curved space geometry.

To substantiate further the intriguing connection between the flat space gauge potential and curved space geometry, consider the curved space metric $g_{\mu\nu}$ which can also be defined in terms of the tetrad field at every point in the space as

$$g_{\mu\nu} = e^i_{\mu} e_{i\nu} = \eta_{ij} e^i_{\mu} e^j_{\nu} \tag{5}$$

where η_{ij} is the local flat space (Lorentzian) metric. As a continuation, it is possible to define the curved space representatives of the gauge potentials as $A^a_{\mu} = e^a_{\mu} A^a_i$ and the field tensor as $F^a_{\mu\nu} = e^i_{\mu} e^j_{\nu} F^a_{ij}$. Consequently, one can construct a fourth rank tensor corresponding to Eq. (4), as (for details see the original paper, Ref. [3]),

$$\tilde{R}_{\mu\nu\rho\lambda} = e^i_{\mu} e^j_{\nu} e^k_{\rho} e^l_{\lambda} \tilde{R}_{ijkl}, \tag{6}$$

which has all the symmetry properties of $\tilde{R}_{ijkl}$. One may immediately be tempted to identify $\tilde{R}_{\mu\nu\rho\lambda}$ with the usual curvature tensor of the Riemannian manifold. But such a correspondence will not be exactly correct. On the other hand, as shown by the authors, it is possible to bifurcate the tensor in to a sum of two as

$$\tilde{R}_{\mu\nu\rho\lambda} = R_{\mu\nu\rho\lambda} + \Sigma_{\mu\nu\rho\lambda} \tag{7}$$

where $R_{\mu\nu\rho\lambda}$ is the curvature tensor of the curved manifold and $\Sigma_{\mu\nu\rho\lambda}$ is a fourth rank tensor in the curved space with the same algebraic properties of the curvature tensor. So, finally the authors have beautifully associated the flat space gauge field to the metric of the curved space and thus find the curvature of the manifold. This is a truly remarkable achievement; firstly, because it does not have the noted shortcoming of the prior theory due to DG; secondly, the theory is not particular about the symmetry group, in fact the theory is true for a large class of symmetry groups.

Further, they have formulated of the field equation of gravity. One can obtain the field equation by varying the Einstein–Hilbert action [4],

$$S_{E-H} = \int \sqrt{-g}\, R\, d^4x \tag{8}$$

where g is the metric scalar and R is the curvature scalar. Following Eq. (6) one can re-write the action as

$$S_{E-H} = \int \sqrt{-g} \left(\tilde{R} - \Sigma \right). \tag{9}$$

This action is essentially a function of the gauge fields and hence can be viewed as the action for the $U(2) \times U(2)$ gauge theory. But this action is not of the YM form. While in the linear limit, where the metric takes the form $g_{\mu\nu} = \eta_{\mu\nu} + h_{\mu\nu}$, the action will exactly reduce to the YM form. Varying this action with respect to $g_{\mu\nu}$, equivalently treating gauge fields as function of the metric components, one will get the second rank curvature tensor as [3],

$$R^{\mu\nu} = -\left[\nabla_i F^{ij,a} \frac{\delta A_j^a}{\delta g_{\mu\nu}} + \nabla_i G^{ij,a} \frac{\delta B_j^a}{\delta g_{\mu\nu}}\right]. \tag{10}$$

This will give the field equation of gravity in free space (in the absence of matter), as $R^{\mu\nu} = 0$.

The remaining task is the coupling with matter field. To a certain extent the authors have succeeded in that where they coupled the curvature with massless Dirac field. By considering matter in the form of the massless Dirac field, they obtained the field equation (please refer to the original article Ref. [3] for details) as

$$R_{\mu\nu} = 3k\left(T_{\mu\nu} - \frac{1}{2} g_{\mu\nu} T_\lambda^\lambda\right) \tag{11}$$

with $k = 8\pi G/3$ and $T_{\mu\nu}$ is the energy-momentum tensor corresponding to the massless Dirac field. The macroscopic effects of gravity thus shown to be dependent on the coefficient k. To explore the microscopic effects of gravitation one has to search further for the details of the coefficients e and f. As a ratification of the formulation, the exact relation between the gauge field and the metric of the curved space has been derived in their article.

4 Conclusion

As believed, the theoretical physics group of Cochin University is actually started under patronage of Prof. Babu Joseph. It was then followed by people like M. Sabir, Ramesh Babu, etc. Prof. Babu Jospeh produced many students and among them, Prof. Sabir was a significant personality [5]. Their union further caused the strengthening of the theory group in the Department of Physics. The referred paper on gravity by Prof. Babu Jospeh and M. Sabir was the first paper in the field of gravity from this group. The paper reformulated Einstein's gravity as a flat space gauge theory by effectively associating the flat space gauge field with the curvature of the curved manifold. Unfortunately, further continuance of this association in the field of gravity would not get the expected momentum due to some reason or other. But, the mentioned paper is truly a remarkable one in the history of the Cochin Theoretical Physics group.

References

1. Yang-Mills theory is basically a gauge theory based on unitary groups to describe the behaviour of the elementary particles. It is used for unifying the week force with electromagnetic force and also to create the quantum theory of the strong force, quantum chromodynamics
2. H. Dehnen, F. Ghaboussi, Nucl. Phys. B **262**, 144; Phys. Rev. D **33**, 2205 (1986)
3. K. Babu Joseph, M. Sabir, Mod. Phys. Lett. A **3**, 497 (1988)

4. T. Padmanabhan, *Gravitation-Foundations and Frontiers* (Cambridge University Press, UK, 2010)
5. Some dates and features of the association between KBJ and MS were gathered from Prof. Sarangadharan, who is a colleague of both

Mathematical Physics

q-Oscillators and a q-deformed Debye Model for Lattice Heat Capacity

K. K. Leelamma

Abstract Quantum groups and quantum algebras are deformations of classical Lie groups and their structure is much more complex than that of Lie groups. They are symmetry groups of non-commutative spaces. The representation theory of the quantum group $SU_q(2)$ has led to the development of q-deformed harmonic oscillator algebra. It has found applications in several branches of physics such as vibrational spectroscopy, nuclear physics, many-body theory and quantum optics. This article introduces the q-oscillator and gives a brief description of the studies carried out by the author using the concept of q-oscillator algebra, under the supervision of Prof. K. Babu Joseph. It includes a q-oscillator Debye model for the lattice heat capacity of crystals and a q-anharmonic oscillator with quartic interaction.

Keywords Quantum group · q-Oscillator · Debye model · q-Commutators · Anharmonic oscillator

1 Introduction

History of physics has several stories of deformations. In classical mechanics itself, the Lorentz transformation between two inertial frames is a deformation of Galilean transformation with $\beta = \frac{v}{c}$ as the deformation parameter. In the limit $\beta \to 0$, the original non-relativistic mechanics is regained. Thus special relativity is a deformation of Galilean relativity. Similarly, quantum mechanics is a deformation of classical mechanics with $\hbar$ as the deformation parameter. In the limit $\hbar \to 0$, the results of quantum mechanics merge with the classical results. Quantum groups and quantum algebras are deformations of classical Lie groups. Quantum group first appeared in physics literature as a deformation of the universal enveloping algebra of a Lie algebra in the study of integrable quantum systems. In the beginning of the 1980s, there was much progress made in the field of quantum integrable field theories. One

K. K. Leelamma (✉)
Department of Physics, Union Christian College, Aluva, Kerala, India
e-mail: leelamma_kk@yahoo.co.in

K. S. Sreelatha and V. Jacob (eds.), *Modern Perspectives in Theoretical Physics*,
https://doi.org/10.1007/978-981-15-9313-0_6

of the most important studies is the development of a quantum mechanical version of the inverse scattering method used in the theory of integrable nonlinear evolution equations like the Korteweg de Vries (KdV) equation. This method was developed by Faddeev et al. [1–3] in formulating a quantum theory of solitons. Kulish and Reshetikhin [4] showed that the quantum linear problem of the quantum sine-Gordon equation was not associated with the Lie algebra $sl(2)$ as in the classical case, but with a deformation of this algebra. Sklyanin [5, 6] showed that deformations of Lie algebraic structures were not special to the quantum sine-Gordon equation and that it seemed to be part of a general theory. It was Drinfel'd who showed that a suitable quantisation of Poisson Lie groups reproduced exactly the same deformed algebraic structures encountered in the theory of quantum inverse scattering [7–9]. Almost at the same time, Jimbo arrived at the same result [10, 11] from a slightly different angle. In his work, the quantum algebras appeared in the context of the solution of the Yang-Baxter Equation (YBE) which is a sufficient condition for solvability of two-dimensional Ising model.

There is no universally accepted definition of a quantum group. There are several approaches. In Drinfeld's approach, the quantum group is defined as a deformation of the universal enveloping algebra of a Lie algebra. This approach is similar to the study of Lie groups via their Lie algebras. Jimbo also gave almost the same definition. The new algebraic structures are called Quantised Universal Enveloping Algebra (QUEA). In Manin's work [12], quantum groups are defined as symmetries of non-commutative or quantum spaces. We discuss this point in detail in Sect. 1.1. Woronowicz [13–15] gave an entirely different approach based on non-commutative C^* algebras. This is analogous to the classical theory of topological groups. He called these groups, pseudo-groups. This approach is popular among mathematicians. The theory of Faddeev and the Leningrad school [16] introduces quantum groups in terms of R-matrices which are solutions of the Quantum Yang-Baxter Equation (QYBE). This approach is directly connected to integrable quantum field theories and has no classical analogue. In all the four approaches, quantum groups have the structure of a Hopf algebra.

Quantum groups and quantum algebras have attracted much attention of physicists and mathematicians during the last decades of the twentieth century, especially after the introduction of the q-harmonic oscillator. The fact that the energy levels of the q-oscillator are not equally spaced and the success of the q-oscillator model in accounting for the measurements on the infrared spectrum of a number of molecules, indicated that q-deformation can take care of anharmonicity effects to some extent. Motivated by these considerations, the Debye model of lattice heat capacity of crystals is reformulated [17], taking each mode as a q-oscillator. In the low temperature limit, the model effectively coincides with the Debye model and in the high temperature limit, C_v is found to be T-dependent, in very good agreement with the experimental results obtained in the three cases studied. This model is discussed in Sect. 2.

The problem of q-deformations of an anharmonic oscillator with quartic interaction [18] and its energy spectrum are studied. The energy values are found to depend on the deformation parameter q. The various thermodynamic quantities such as partition function, entropy and internal energy are also evaluated. It is discussed in Sect. 3. These studies are expected to be of relevance in the context of lattice dynamics.

The concept of q-deformation is also applied to investigate the magnetic properties of ferromagnets [19]. The agreement between the linear spin-wave theory of ferromagnetism and experimental observations on ferromagnets is not satisfactory. The q-deformed Holstein-Primakoff transformation is used to describe the spin variables of a Heisenberg ferromagnet and the magnons are treated as q-bosons. The exchange Hamiltonian in the nearest neighbour approximation is obtained for small values of the deformation parameter when the excitation is low. The thermodynamic quantities in the low temperature region are also evaluated. It is found that the spontaneous magnetisation and magnetic contribution to the heat capacity have q-dependent $T^{\frac{1}{2}}$ terms in addition to the well-known Bloch $T^{\frac{3}{2}}$ term. A comparison of the theoretical results and experimental values in the cases of EuO and EuS, the simplest Heisenberg ferromagnets, indicated that our model is an improvement over the linear spin-wave theory.

1.1 Quantum Groups and non-commutative Spaces

The quantum algebras have been linked to geometries that have non-commutative structures [20, 21]. The concept of space-time continuum has been fundamental to all successful physical theories. However there are arguments that on a sub-microscopic level, this concept has to be abandoned [22]. There is no experimental proof for the assumption that space-time is smooth down to arbitrarily small distances. Perhaps it may be because of the idealisation of space-time concept that one comes across tremendous problems in the unification of various interactions [23]. This motivates one to look for a new space-time concept. Quantum mechanical phase-space is only partially non-commuting, only co-ordinates and momenta non-commute, co-ordinates themselves are commuting. If at a sufficiently small length scale, co-ordinates become non-commuting operators, it will be impossible to measure the position of a particle exactly. In this way, one may hope to remove the ultraviolet divergences of conventional quantum field theory which are due to the possibility of measuring field oscillations at one point. Thus non-commutativity is introduced as a necessary condition in the generalised space-time concept. It has been argued that physics at the Planck scale may be understood only with the help of non-commutative geometry [23, 24].

In a non-commutative space with real co-ordinates (x, y, z), a unit of length along the x-direction is defined as

$$\Delta x = (q - 1)x \tag{1}$$

or equivalently

$$\Delta x = (q - q^{-1})x \tag{2}$$

where q is some parameter which is real. The width of the interval Δx is not a constant. In the limit $q \to 1$, the interval $\Delta x \to 0$ and we have space-time continuum.

Consider a system with two degrees of freedom. The quantum mechanical phase-space of the system is spanned by the co-ordinates x, y and conjugate momenta p_x and p_y. The phase-space is only partially non-commuting:

$$[x, p_x] = i\hbar = [y, p_y] \tag{3}$$

$$[x, y] = 0 = [p_x, p_y] \tag{4}$$

i.e., the $x - y$ plane and $p_x - p_y$ plane have continuum structure and only $x - p_x$ and $y - p_y$ planes may have discrete structure. In a non-commutative space, non-commutativity is prescribed for co-ordinates also:

$$xy = qyx \tag{5}$$

The q-commutator

$$[x, y]_q = xy - qyx = 0 \tag{6}$$

In general, for any two operators A and B, $[A, B]_q = AB - qBA$, where q is some parameter which may be real or complex. It satisfies the properties $[A, B]_q = -q[B, A]_{q^{-1}}$ and $\lim_{q\to1}[A, B]_q = [A, B]$, the usual commutator in quantum mechanics. The q-commutators do not satisfy the Jacobi identity.

Equation (5) should remain covariant under a co-ordinate transformation $(x, y) \to (x', y')$.
Let

$$T = \begin{pmatrix} a & b \\ c & d \end{pmatrix} \tag{7}$$

be the matrix effecting the transformation. a, b, c and d are in general non-commuting elements. Then

$$\begin{pmatrix} x' \\ y' \end{pmatrix} = \begin{pmatrix} a & b \\ c & d \end{pmatrix} \begin{pmatrix} x \\ y \end{pmatrix} = \begin{pmatrix} ax + by \\ cx + dy \end{pmatrix}$$

$x'y' = qy'x'$ implies

$$(ax + by)(cx + dy) = q(cx + dy)(ax + by) \tag{8}$$

If we assume that a, b, c, d commute with (x, y) we can write

$$\begin{pmatrix} x' & y' \end{pmatrix} = \begin{pmatrix} x & y \end{pmatrix} \begin{pmatrix} a & b \\ c & d \end{pmatrix} = \begin{pmatrix} ax + cy & bx + dy \end{pmatrix}$$

The invariance of Eq. (5) implies

$$(ax + cy)(bx + dy) = q(bx + dy)(ax + cy) \tag{9}$$

Equations (8) and (9) give a complete set of conditions to be obeyed by the non-commuting objects a, b, c, d to preserve the structure of the quantum plane:

$$ab = qba;\ cd = qdc;\ ac = qca;\ bd = qdb;\ bc = cb;$$

$$ad - da = (q - q^{-1})bc \tag{10}$$

These are commutation relations obeyed by a, b, c, d. T is called a quantum matrix. It is shown that [25] the matrices T satisfy all the axioms of a non-commutative Hopf algebra and thus constitute a quantum group. It is denoted by $GLq(2)$, the quantum linear general group in two dimensions. It is the group of linear transformations in two-dimensional non-commutative space that preserves the commutation relation (5). The additional relation

$$ad - qbc = 1 \tag{11}$$

yields the quantum unimodular group $SLq(2)$. The object defined by

$$det_q(T) = ad - qbc \tag{12}$$

is called the quantum determinant or q-determinant. Relation (5) can be generalised to the case of two or more copies of quantum planes [26].

1.2 q-Deformed Numbers and q-Differential Calculus

The q-deformation of numbers was introduced by Heine [27] in 1878. The q-number $[n]_q$ corresponding to the ordinary number n is defined as

$$[n]_q = \frac{q^n - q^{-n}}{q - q^{-1}} \tag{13}$$

This definition of q-deformation possesses $q \leftrightarrow q^{-1}$ symmetry

$$lim_{q\to 1}[n]_q = n \tag{14}$$

The q-functions are also defined [28]. For example, the q-exponential function

$$e_q(x) = \sum_{n=0}^{\infty} \frac{x^n}{[n]_q!} \tag{15}$$

where the q-factorial

$$[n]_q! = [n]_q[n-1]_q....[2]_q[1]_q \tag{16}$$

It follows that

$$[1]_q = 1; \quad [0]_q = 0; \quad [0]_q! = 1 \tag{17}$$

The q-sine and q-cosine functions are defined as

$$sin_q(x) = \frac{1}{2i}(e_q(ix) - e_q(-ix)) \tag{18}$$

$$cos_q(x) = \frac{1}{2}(e_q(ix) + e_q(-ix)) \tag{19}$$

q-differential calculus is a generalisation of ordinary differential calculus. It was developed in the nineteenth century by Jackson [29, 30]. Let $f(x)$ be a function of the real variable x. Its q-derivative is defined as

$$D_x f(x) = \frac{f(qx) - f(q^{-1}x)}{x(q - q^{-1})} \tag{20}$$

where q is in general some complex parameter. The q-derivative becomes the ordinary derivative as $q \to 1$.

$$lim_{q\to 1} D_x(f) = \frac{\partial f}{\partial x} = \partial_x f \tag{21}$$

Thus q-differentiation defines a finite differential calculus where the intervals are finite. As $q \to 1$, $\Delta x \to 0$ and the variation of x is continuous. In this respect, q-differential calculus is convenient for the description of non-commutative space. The q-derivative satisfies the following properties:

$$D_x(x) = 1 \tag{22}$$

$$D_x(x^n) = [n]_q x^{n-1} \tag{23}$$

$$D_x(ax^n) = a[n]_q x^{n-1} \tag{24}$$

$$D_x(af + bg) = aD_x(f) + bD_x(g) \tag{25}$$

$$D_x(fg) = g(x)D_x(f) + f(qx)D_x(g) \tag{26}$$

$$D_x x - q^{-1} x D_x = q^{x\partial_x} \tag{27}$$

$$[x\partial_x, x] = x \tag{28}$$

$$[x\partial_x, D_x] = -D_x \tag{29}$$

Here a and b are constants and f and g are functions of x.

1.3 $su_q(2)$ Algebra

One of the well-studied quantum groups is $SU_q(2)$ which is the q-deformation of the classical group $SU(2)$, the group of angular momentum. The Lie algebra $su(2)$ consists of three elements L_+, L_- and L_z which satisfy the commutation relations

$$[L_z, L_\pm] = \pm L_\pm \tag{30}$$

$$[L_+, L_-] = 2L_z \tag{31}$$

with

$$L_+^\dagger = L_- \tag{32}$$

Kulish and Reshetikhin [4], while studying the solution of YBE, introduced the algebra of three elements J_+, J_- and J_z:

$$[J_z, J_\pm] = \pm J_\pm;$$

$$[J_+, J_-] = [2J_z]_q \tag{33}$$

In the limit $q \to 1$, this algebra goes into $su(2)$ algebra. Thus it is called q-deformation of $su(2)$ algebra and is denoted by $su_q(2)$. Both $su_q(2)$ and $sl_q(2)$ are quantum algebras with a single deformation parameter q. Going to higher dimensions with more than two non-commuting co-ordinates, one has to use more than one deformation parameter. Several authors have worked on two-parameter deformations [31, 32].

1.4 q-Harmonic Oscillator

The simple harmonic oscillator (SHO) problem has an indispensable role in physics. It is customary to use the SHO to illustrate the basic concepts and new methods in classical as well as quantum physics. The wave mechanical theory of oscillators provides the basis for understanding the properties of a wide variety of systems which are analysable in terms of harmonic oscillators. It is useful not only in the study of vibrations of diatomic and polyatomic molecules, but also in the study of vibrations of more complicated systems expressed in terms of their normal modes. Thus its applications are not limited to molecular spectroscopy, but extend to a variety of branches of modern physics such as condensed matter physics, nuclear structure, quantum field theory, quantum optics, quantum statistical mechanics and so forth. The studies carried out by the author is essentially based on the q-deformed harmonic oscillator introduced in 1989 by Biedenharn [33] and independently by Macfarlane [34].

The creation, annihilation and number operators $a^\dagger$, a and N $(= a^\dagger a)$ satisfy the commutation relations

$$[a, a^\dagger] = 1 \tag{34}$$

$$[N, a^\dagger] = a^\dagger \tag{35}$$

$$[N, a] = -a \tag{36}$$

The generators of $su(2)$ can be realised in the form

$$J_+ = -\frac{1}{2}a^{\dagger 2}; \quad J_- = \frac{1}{2}a^2; \; J_z = \frac{1}{4}(aa^\dagger + a^\dagger a) \tag{37}$$

To define a q-analogue to the harmonic oscillator, the q-deformed operators a_q and $a_q^\dagger$, and the q-boson vacuum ket $|0\rangle_q$ are considered. Biedenharn and Macfarlane independently found two-commutation relations for a_q and $a_q^\dagger$, which are equivalent and which can be written in an alternate form [35] as

$$[a_q, a_q^\dagger]_q = a_q a_q^\dagger - q a_q^\dagger a_q = q^{-N_q} \tag{38}$$

where the operator N_q satisfies the commutation relations

$$[N_q, a_q^\dagger] = a_q^\dagger; \quad [N_q, a_q] = -a_q \tag{39}$$

N_q is Hermitian, but

$$N_q \neq a_q^\dagger a_q \tag{40}$$

$[a_q, a_q^\dagger]_q$ is the q-commutator. The algebra defined by (38) and (39) is referred to as q-oscillator algebra. In the limit $q \to 1$, it tends to the standard oscillator algebra. The operators $a_q^\dagger$, a_q and N_q are referred to as q-boson creation operator, q-boson annihilation operator and q-boson number operator, respectively.

To realise the Lie algebra of the generators of $SU_q(2)$, a pair of mutually commuting q-harmonic oscillator systems with operators a_{iq} and $a_{iq}^\dagger$ with $i = 1, 2$ is considered. Then the q-analogue of the Jordan-Schwinger map is defined:

$$J_+ = a_{1q}^\dagger a_{2q}; \quad J_- = a_{2q}^\dagger a_{1q} = J_+^\dagger; \quad J_z = \frac{1}{2}(N_{1q} - N_{2q}) \tag{41}$$

These generators satisfy the $su_q(2)$ algebra (33).

1.4.1 Properties of q-Boson Operators

From the defining relations (38) and (39), the following properties of q-boson operators can be deduced:

(1)

$$a_q^\dagger f(N_q) = f(N_q - 1)a_q^\dagger \tag{42}$$

$$a_q f(N_q) = f(N_q + 1)a_q \tag{43}$$

$$[N_q, a_q^\dagger a_q] = [N_q, a_q a_q^\dagger] = 0 \tag{44}$$

or in general

$$[a_q a_q^\dagger, f(N_q)] = [a_q^\dagger a_q, f(N_q)] = 0 \tag{45}$$

Here $f(N_q)$ is an arbitrary function of N_q.

(2) The parameter q is either real or is a pure phase. If q is real, it can be expressed as

$$q = e^\eta \tag{46}$$

η being real. Then the q-number

$$[x]_q = \frac{q^x - q^{-x}}{q - q^{-1}} = \frac{e^{\eta x} - e^{-\eta x}}{e^\eta - e^{-\eta}} = \frac{\sinh(\eta x)}{\sinh(\eta)} \tag{47}$$

If q is a pure phase, of the form

$$q = e^{i\eta} \tag{48}$$

$$[x]_q = \frac{e^{i\eta x} - e^{-i\eta x}}{e^{i\eta} - e^{-i\eta}} = \frac{\sin(\eta x)}{\sin(\eta)}. \tag{49}$$

(3) The bilinear forms become

$$a_q^\dagger a_q = [N_q]_q \tag{50}$$

$$a_q a_q^\dagger = [N_q + 1]_q \tag{51}$$

(4) The above properties exhibit $q \leftrightarrow q^{-1}$ symmetry.

(5) In the limit $q \to 1$, the q-numbers (or operators) tend to the ordinary numbers (or operators).

$$\lim_{q\to 1}[x]_q = \lim_{\eta\to 0} \frac{sin(\eta x)}{sin(\eta)} = x$$
$$\lim_{q\to 1}(a_q^\dagger a_q) = \lim_{q\to 1}[N_q]_q = N_q$$
$$\lim_{q\to 1}(a_q a_q^\dagger) = \lim_{q\to 1}[N_q + 1]_q = N_q + 1$$

(6) In general, $N_q \neq a_q^\dagger a_q$. However, Polychronakos [36] has shown that there exists a classical realisation of the q-oscillator algebra in which the new generators are defined as

$$a_q^\dagger = \sqrt{\frac{f(N)}{N}} a^\dagger \tag{52}$$

$$a_q = \sqrt{\frac{f(N+1)}{N+1}} a \tag{53}$$

and

$$N_q = N = a^\dagger a \tag{54}$$

where $f(N) = \frac{q^N - q^{-N}}{q - q^{-1}}$. In this,

$$a_q a_q^\dagger - q a_q^\dagger a_q = q^{-N} \tag{55}$$

Then

$$a_q^\dagger a_q = [N]; \quad a_q a_q^\dagger = [N+1] \tag{56}$$

We call this as the *boson realisation* of the q-oscillator algebra. In this realisation, the eigenstates $|n\rangle_q$ of N_q are the same as those of the usual harmonic oscillator:

$$N_q \, |n\rangle_q = N \, |n\rangle_q = n \, |n\rangle_q \tag{57}$$

i.e., the eigenvalues of N_q are also the integers from 0 to ∞, and hence N_q is interpreted as the number of q-deformed bosons. It is postulated that there exists a vector $|0\rangle_q$ with the properties

$$a_q \, |0\rangle_q = a_q \, |0\rangle = 0; \quad N_q \, |0\rangle_q = N_q \, |0\rangle = 0 \tag{58}$$

$|0\rangle_q$ is referred to as the q-deformed vacuum state and is interpreted as a state without bosons. The interpretation of $a_q^\dagger$ and a_q as the creation and annihilation operators also holds. The eigenstates $|n\rangle_q$ are orthonormal.

$$|n\rangle_q = \frac{(a_q^\dagger)^n}{\sqrt{[n]_q!}} |0\rangle_q \tag{59}$$

Also

$$a_q |n\rangle_q = \sqrt{[n]_q} |n-1\rangle_q \tag{60}$$

$$a_q^\dagger |n\rangle_q = \sqrt{[n+1]_q} |n+1\rangle_q \tag{61}$$

$$_q\langle n+1|a_q^\dagger|n\rangle_q = \sqrt{[n+1]_q} \tag{62}$$

$$_q\langle n-1|a_q|n\rangle_q = \sqrt{[n]_q} \tag{63}$$

The Hilbert space spanned by $\{|n\rangle_q\}$ is positive definite only if $|q| \leq 1$. For larger values of $|q|$, states with negative squared norm arise and the probability interpretation of quantum mechanics is lost.

1.4.2 Energy Spectrum of q-deformed Harmonic Oscillator

The Hamiltonian of the q-deformed harmonic oscillator is

$$H_q = \frac{p_q^2}{2m} + \frac{1}{2} m\omega^2 x_q^2 \tag{64}$$

where the q-position x_q and the q-momentum p_q of the oscillator are related to the q-creation and q-annihilation operators $a_q^\dagger$ and a_q as

$$x_q = \sqrt{\frac{\hbar}{2m\omega}}(a_q^\dagger + a_q) \tag{65}$$

$$p_q = i\sqrt{\frac{m\hbar\omega}{2}}(a_q^\dagger - a_q) \tag{66}$$

where a_q and $a_q^\dagger$ satisfy the q-oscillator algebra (38) and (39). Then the Hamiltonian reads

$$H_q = \frac{1}{2}\hbar\omega(a_q^\dagger a_q + a_q a_q^\dagger) = \frac{1}{2}\hbar\omega([N]_q + [N+1]_q) \tag{67}$$

Here we are using the boson realisation of the q-oscillator algebra in which $N_q = N$. The number and energy eigenstates of the q-oscillator are then the same as those of the usual harmonic oscillator and are q-independent. Only eigenvalues are q-dependent. The energy eigenvalues are given by

$$E_{qn} = \frac{1}{2}\hbar\omega([n]_q + [n+1]_q) \tag{68}$$

i.e., the energy levels of the q-oscillator are not uniformly spaced for $q \neq 1$. The behaviour of the energy spectra is completely different in the cases $q = e^{\eta}$ and $q = e^{i\eta}$. When q is real ($q = e^{\eta}$), the separation between the levels increases with the value of n. On the other hand, when q is a pure phase, the separation between the levels decreases with increasing n. The spectrum in this case exhibits many characteristic features of the anharmonic oscillator. The energy levels of the anharmonic oscillator are not equidistant, but their separation decreases as the value of the oscillator quantum number v increases:

$$E_v = \hbar\omega(v + \frac{1}{2}) - \hbar\omega x_e(v + \frac{1}{2})^2 + \hbar\omega y_e(v + \frac{1}{2})^3 - \cdots \tag{69}$$

where $\omega y_e << \omega x_e << \omega$ and v takes only a limited number of values ($v \leq v_{max}$) because of the finite depth of the potential well. The energy values of the q-deformed harmonic oscillator can be written as

$$E_{qn} = \frac{1}{2}\hbar\omega([n]_q + [n+1]_q)$$

$$= \begin{cases} \frac{1}{2}\hbar\omega\left(\frac{\sinh\eta(n+\frac{1}{2})}{\sinh\frac{\eta}{2}}\right) \text{ if } q = e^{\eta} \\ \frac{1}{2}\hbar\omega\left(\frac{\sin\eta(n+\frac{1}{2})}{\sin\frac{\eta}{2}}\right) \text{ if } q = e^{i\eta} \end{cases} \tag{70}$$

On expanding the second expression, we get

$$E_{qn} = \frac{1}{2}\hbar\omega\frac{\eta}{\sin(\frac{\eta}{2})}\{(n + \frac{1}{2}) - \frac{\eta^2}{6}\left(n + \frac{1}{2}\right)^3 + \cdots\} \tag{71}$$

Comparing this with the expression (69), we see that there is great similarity between the spectrum of the q-deformed harmonic oscillator and that of the anharmonic oscillator describing the vibrational spectra of diatomic molecules. However, the coincidence is only a qualitative one. Expression (71) contains only the odd powers of $(n + \frac{1}{2})$ whereas expression (69) contains odd as well as even powers of $(v + \frac{1}{2})$.

$$\lim_{q\to 1 or \eta\to 0} E_{qn} = \hbar\omega\left(n + \frac{1}{2}\right) \tag{72}$$

i.e., in the limit $q \to 1$, the energy spectrum of q-deformed harmonic oscillator coincides with that of the standard harmonic oscillator. Besides the energy spectrum, other properties of q-oscillators are also well studied. For example, coherent states and squeezed states of q-harmonic oscillators have been investigated by Vinod et al. [37].

1.4.3 Studies of Physical Systems using q-Oscillator Algebra

Numerous applications of quantised algebra to real physical systems have been worked out by various authors. A few of them are cited here. When used to describe the vibrational spectra of diatomic molecules [38] such as H_2, it is seen that when q is chosen as a pure phase, the results show fair agreement with the experimental data, for $\eta \simeq 0.06$. A q-rotator model with $su_q(2)$ symmetry has been set up to describe the rotational spectra of diatomic molecules [39]. For deformation parameter $\eta \simeq 0.01$, the spectra of the q-rotator model coincide with the observed spectra to satisfactory accuracy. A complete quantum group theoretic treatment of vibrating and rotating diatomic molecules has also been given [40] by assuming the deformation parameter q to depend on the rotational quantum number J. The coincidence between the predictions of the model and conventional phenomenological formulae is remarkable. The $su_q(2)$ algebra has been used for the description of energy spectra of the deformed even-even nuclei [41], and it is shown that there is good agreement with experimental results when q is chosen as a phase with $\eta \simeq 0.04$. The many-body problem of q-oscillators has been investigated by several authors [42–44]. The spectra of the system are found to be rich, exhibiting interactions between the levels of the individual oscillators. The deformed algebra has also been employed to the many-body problem of composite particles. The q-oscillator models in two and higher dimensions are applied to the spectra of triatomic molecules such as H_2O and superdeformed nuclei. The nature of an electromagnetic field of high intensity modelled by q-oscillators also has been studied [45].

2 q-Oscillator Debye Model for Lattice Heat Capacity

The Debye model for lattice heat capacity is modified retaining all the basic assumptions except that each mode is here treated as a q-deformed harmonic oscillator [17]. The two basic experimental facts about the heat capacity of solids which any theory must explain are (i) at room temperature, the heat capacity of most solids is close to $3k_B$ per atom so that for molecules consisting of n atoms, the molar heat capacity is close to $3nR$ where R is the universal gas constant. Accurate measurements indicate temperature dependence of heat capacity in this region. (ii) At low temperatures, the heat capacities decrease and vanish at $T = 0$ K. The decrease goes as T^3.

The Debye model has been successful in describing the experimental observations at low temperatures in many pure crystalline solids. In the low temperature regime, the Debye's theory predicts

$$C_V = \frac{12\pi^4}{5} N_0 k_B \left(\frac{T}{\theta_D}\right)^3 \quad (73)$$

where N_0 is the total number of atoms in the crystal and θ_D is the Debye temperature of the solid. Thus $C_v \propto T^3$ in agreement with experimental results. In the high temperature region $T >> \theta_D$, the Debye model leads to the Dulong-Petit law:

$$C_v = 3R/g.atom \quad (74)$$

a constant for all monoatomic crystals and is independent of temperature. This is not in exact agreement with experimental observations which show an increase of heat capacity with temperature.

Debye's theory involves three basic assumptions: (i) isotropy of the solid (ii) nondispersion of sound waves in the medium and (iii) degeneracy of different branches of allowed modes. Above all, it is based on the harmonic approximation. Real crystals do exhibit anharmonic effects such as thermal expansion; the adiabatic and isothermal elastic constants are in general different and dependent on temperature and pressure. The influence of the anharmonicity on the various quantities for specific cases has been dealt with in a number of papers see, for example, [46]. Motivated by the fact that q-deformation can take care of anharmonicity effects to some extent, it is tried to explain the temperature dependence of lattice heat capacity in the temperature region $T >> \theta_D$ by suggesting a q-oscillator Debye model.

The properties of q-deformed harmonic oscillators are discussed at length in Sect. 1.3. In the present model, the phonon modes are treated as slightly deformed q-oscillators (SDO) in the boson realisation. That is, η is taken to be very small, close to zero so that only terms upto $O(\eta^2)$ are retained in

$$H_q = \frac{1}{2}\hbar\omega([N+1]+[N])$$

Hamiltonian of the slightly deformed oscillator is obtained as

$$H_{SDO} = H_0 - \frac{\eta^2}{3!} H_1 \quad (75)$$

where H_0 is the Hamiltonian of the usual harmonic oscillator:

$$H_0 = \frac{1}{2}\hbar\omega(2N+1) \quad (76)$$

$$H_1 = \frac{1}{2}\hbar\omega[(N+1)^3 + N^3 - (2N+1)] \quad (77)$$

The partition function and internal energy of the slightly deformed harmonic crystal are evaluated and the lattice heat capacity C_v of the crystal is computed in two limiting cases:

(a) T $<<\theta_D$:

$$C_v = \frac{12\pi^4}{5} N_0 k_B \left(\frac{T}{\theta_D}\right)^3 \left(1 + \frac{45}{2}\frac{\eta^2}{\pi^2}\right) \tag{78}$$

Comparing this with Eq. (73), we see that q-deformation brings in a q-dependent correction which is negligible. Thus the model coincides with the Debye model in the low temperature limit.

(b) T $>>\theta_D$:

$$C_v = 3N_0 k_B \left(1 + \eta^2 \frac{18T^2}{\theta_D^2}\right) \tag{79}$$

$= 3R(1 + \eta^2 \frac{18T^2}{\theta_D^2})$/per g. atom for a monoatomic solid.

This expression exhibits a T^2 dependence in contrast to Eq. (74). The lattice heat capacity per g.atom is calculated according to the above expression for the three alkali elements Potassium, Rubidium and Caesium for which the Debye temperatures are relatively low. η is assigned values $\sim 10^{-2}$. The results are plotted in Fig. 1 for the range 100–300 K along with the experimental values [47, 48].

It is observed that there is very good agreement for not too high values of T. As the temperature becomes higher, discrepancies arise, the heat capacity increases much more rapidly than that predicted by the theory. The above investigations lend support to the view that phonons in crystals may be q-quantised excitations. The deviations observed at higher temperatures may be explained taking into account quartic and higher order interactions possibly within the framework of a q-anharmonic oscillator model.

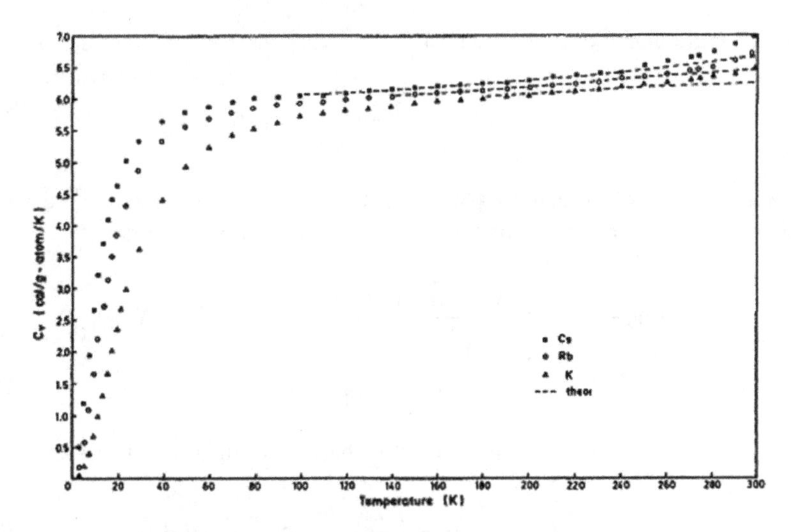

Fig. 1 Values of lattice heat capacity calculated in the q-harmonic approximation plotted as a function of temperature for alkali metals Cs, Rb and K. Experimental values are also shown

3 q-Anharmonic Oscillator (q-AO) with quartic Interaction

The q-deformation of an anharmonic oscillator with quartic interaction is studied in the first-order perturbation theory and the energy spectrum is investigated using the boson realisation of the unperturbed q-oscillator eigenstates. The partition function, entropy and free energy are also evaluated [18].

The Hamiltonian of the q-AO is taken to be

$$\bar{H} = \frac{\bar{p}^2}{2m} + \frac{1}{2}m\omega^2\bar{x}^2 + \frac{\lambda}{4!}\bar{x}^4 \tag{80}$$

The q-position operator $\bar{x}$ and the q-momentum operator $\bar{p}$ of the q-AO are related to the q-boson operators a_q and $a_q^\dagger$ in the same way as in the case of the q-harmonic oscillators. We work in the boson realisation in which $N_q = N = a^\dagger$ a and the eigenstates are those of the usual harmonic oscillator. Hereafter we drop the suffix q for q-deformed operators and q-numbers for convenience. Thus

$$\bar{x} = \frac{\hbar}{2m\omega}(a + a^\dagger) \tag{81}$$

$$\bar{p} = i\frac{\hbar m\omega}{2}(a - a^\dagger) \tag{82}$$

where

$$[N, a^\dagger] = a^\dagger; \quad [N, a] = -a; \quad aa^\dagger - qa^\dagger a = q^{-N} \tag{83}$$

The Hamiltonian then takes the form

$$\bar{H} = \frac{1}{2}\hbar\omega(a^\dagger a + aa^\dagger) + \frac{\lambda}{4!}\left(\frac{\hbar}{2m\omega}\right)^2 (a + a^\dagger)^4 \tag{84}$$

Retaining only those terms in $(a + a^\dagger)^4$ yielding nonzero contribution to the expectation values and using the properties (42) and (43) of q-boson operators,

$$\begin{aligned}\bar{H} = \frac{1}{2}\hbar\omega([N+1] + [N]) + \frac{\lambda}{4!}\left(\frac{\hbar}{2m\omega}\right)^2 \Big\{&[N+1][N+2] + [N+1][N+1] \\ &+ 2[N+1][N] + [N][N] + [N][N-1]\Big\}\end{aligned} \tag{85}$$

In the limit q $\rightarrow$ 1 the q-number operators become ordinary operators and

$$\bar{H} = \hbar\omega\left(N + \frac{1}{2}\right) + \frac{\lambda}{4!}\left(\frac{\hbar}{2m\omega}\right)^2 (6N^2 + 6N + 3) \tag{86}$$

which is the same as the Hamiltonian of the ordinary anharmonic oscillator. To get an explicit expression for the Hamiltonian $\bar{H}$ of the q-AO, we consider only slight deformations for which q is very close to unity or η is very close to zero. Also q is chosen to be real of the form $q = e^{\eta}$. Then

$$[N] = \frac{\sinh(N\eta)}{\sinh(\eta)} \tag{87}$$

The hyperbolic functions are expanded in Taylor series in powers of η and we retain only terms upto $O(\eta^2)$ in $\bar{H}$. The Hamiltonian for the slightly deformed anharmonic oscillator (SDAO) is obtained as

$$\begin{aligned}
\bar{H}_{SDAO} &= \frac{1}{2}\hbar\omega(2N+1) + \frac{\eta^2}{3!}\frac{1}{2}\hbar\omega((N+1)^3 + N^3 - (2N+1)) \\
&+ \frac{\lambda}{4!}\left(\frac{\hbar}{2m\omega}\right)^2 (6N^2 + 6N + 3) \\
&+ \frac{\eta^2}{3!}\frac{\lambda}{4!}\left(\frac{\hbar}{2m\omega}\right)^2 (12N^4 + 24N^3 + 36N^2 + 24N + 6) \\
&= H_0 + \frac{\eta^2}{3!}H_1 + H' + \frac{\eta^2}{3!}H''
\end{aligned} \tag{88}$$

As $q \to 1$, $\eta \to 0$ and the above expression tends to (86), the Hamiltonian of the usual boson anharmonic oscillator. The quartic anharmonic corrections (to first order in λ) to the energy levels of the SDAO follow at once by calculating $\langle n|\bar{H}_{SDAO}|n\rangle$ where $|n\rangle' s$ are the unperturbed eigenstates. The partition function for the SDAO is then obtained as

$$Z_{SDAO} = Tr.\exp(-\beta\bar{H}_{SDAO}) \approx Z_0\left\{1 - \beta\left(\frac{\eta^2}{3!}\langle H_1\rangle + \langle H'\rangle\right)\right\} \tag{89}$$

where

$$Z_0 = \Sigma_n\langle n|\exp(-\beta H_0)|n\rangle \tag{90}$$

is the partition function of the ordinary harmonic oscillator and $\langle H_1\rangle$ and $\langle H'\rangle$ are the thermal averages of H_1 and H' respectively. It is assumed that η and λ are very small so that the last term in $\bar{H}_{SDAO}$ which contains both η^2 and λ is neglected. The free energy F, internal energy U and entropy S of the $SDAO$ are also evaluated [18]. It is found that the expressions for F, U and S consist of q dependent correction terms and in the limit $q \to 1$, the results coincide with the classical results.

The relevance of the study of the q-anharmonic oscillator and its thermodynamics will be clear only when it is applied to some real physical systems. A possible scenario for further investigation is lattice dynamics.

4 Conclusion

Quantum algebras, being nonlinear extension of Lie algebras, are specially suited for describing small perturbations in systems characterised by Lie symmetries. Their use in physics became popular with the introduction of the q-deformed harmonic oscillator as a tool for providing a boson realisation of the quantum algebra $su_q(2)$. The q-oscillator algebra has been proved to be useful for the description of small deviations in the behaviour of physical systems from that predicted by harmonic approximation. Several examples for this have been mentioned. During the last three decades, a great deal of work has been done in physics and mathematics, with different approaches providing new view points leading to surprising possibilities such as q-deformed theories of gravitation.

References

1. L.D. Faddeev, *Integrable Models in (1+1) Dimensional Quantum Field Theory*, Les Houches session XXXIX (1982), p. 563
2. L.D. Faddeev, E.K. Sklyanin, L.A. Takhtajan, Theor. Math. Phys. **40**, 194 (1979)
3. L.D. Faddeev, L.A. Takhtajan, Russ. Math. Surv. **34**(5), 11 (1979)
4. P.P. Kulish, N.Y. Reshetikhin, J. Soviet. Math. **23**, 2435 (1983)
5. E.K. Sklyanin, Func. Anal. Appl. **16**, 263 (1982)
6. E.K. Sklyanin, Func. Anal. Appl. **17**, 263 (1982)
7. V.G. Drinfel'd, Sov. Math. Dokl. **32**, 254 (1985)
8. V.G. Drinfel'd, Quantum groups, in *Proceedings of the International Congress of Mathematicians*, Berkeley (1986); Am. Math. Soc. **798** (1987)
9. V.G. Drinfel'd, J. Sov. Math. **41**, 18 (1988)
10. M. Jimbo, Lett. Math. Phys. **10**, 63 (1985)
11. M. Jimbo, Lett. Math. Phys. **11**, 247 (1986)
12. Y.I. Manin, *Quantum Groups and Non-commutative Geometry*, Preprint Montreal Univ. CRM-1561 (1988)
13. S.L. Woronowicz, Comm. Math. Phys. **111**, 613 (1987)
14. S.L. Woronowicz, Invent. Math. **93**, 35 (1988)
15. S.L. Woronowicz, Comm. Math. Phys. **122**, 125 (1989)
16. L.D. Faddeev, N.Y. Reshetikhin, L.A. Takhtajan, *Quantisation of Lie Groups and Lie Algebras*, LOMI Preprint E-14-87
17. K.K. Leelamma, V.C. Kuriakose, K. Babu Joseph, Int. J. Mod. Phys. B **7**, 2697(1993)
18. V.C. Kuriakose, K.K. Leelamma, K. Babu Joseph, Pramana J. Phys. **39**, 521 (1992)
19. K.K. Leelamma, *Studies in Condensed Matter Physics Using q-Oscillator Algebra*, chapter 5, Ph.D. thesis submitted to CUSAT, Kochi (1997)
20. J. Wess, B. Zumino, Nucl. Phys. B Proc. Suppl. **18B**, 302 (1991)
21. B. Zumino, Mod. Phys. Lett. A **6**, 1225 (1991)
22. S. Majid, J. Class. Quantum Gravity **5**, 1587 (1988)
23. F.M. Hoissen, J. Phys. A: Math. Gen. **25**, 1703 (1992)
24. Y. Aref'eva, I.V. Volovich, Preprint CERN-TH 6137/91 (1991)
25. R. Chakrabarthi, R. Jagannathan, J. Phys. A **24**, 5683 (1991)
26. J. Wess, B. Zumino, Nucl. Phys. B (Proc. Suppl.) **18B** 302 (1990)
27. E. Heine, *Handbuch der Kugelfunktionen* (Reamer, Berlin, 1878) (reprinted by Physica-Verlag, Wurzburg, vol. 1, 1961)

28. H. Exton, *q-Hypergeometric Functions and Applications* (Ellis Horwood, Chichester, 1983)
29. F. Jackson, Trans. R. Soc. **46**, 1253 (1908)
30. F. Jackson, Q. J. Math. **41**, 193 (1910)
31. R. Jagannathan, R. Sridhar, R. Vasudevan, S. Chaturvedi, M. Krishnakumari, P. Shanta, V. Sreenivasan, J. Phys. A **25**, 6429 (1992)
32. R. Chakrabarti, R. Jagannathan, J. Phys. A: Math. Gen. **24**, L711 (1991)
33. L.C. Biedenharn, J. Phys. A **22**, L 873 (1989)
34. A.J. Macfarlane, J. Phys. A **22**, L 4581 (1989)
35. M. Chaichan, P. Kulish, Phys. Lett. B **234**, 72 (1990)
36. A.P. Polychronakos, Mod. Phys. Lett. A **5**, 2325 (1990)
37. G. Vinod, K. Babu Joseph, V.C. Kuriakose, Pramana J. Phys. **42**, 299 (1994)
38. D. Bonatsos, E.N. Argyres, P. Raychev, J. Phys. A: Math. Gen. **24**, L403 (1991)
39. Z. Chang, H. Yan, Phys. Lett. A **154**, 254 (1991)
40. Z. Chang, H. Yan, Phys. Lett. A **158**, 242 (1991)
41. P. Raychev, R.P. Roussev, Y.F. Smirnov, J. Phys. G: Nucl. Part. Phys. **16**, L137 (1990)
42. E.G. Floratos, J. Phys. A: Math. Gen. **24**, 4739 (1991)
43. M. Chaichian, R. Gonzale Felipe, C. Montonen, J. Phys. A: Math. Gen. **26**, 4025 (1993)
44. A.K. Mishra, G. Rajasekharan, Pramana J. Phys. **45**, 91 (1995)
45. V.I. Man'ko, G. Marmo, F. Zaccaria, Phys. Lett. A **191**, 13 (1994)
46. P.C. Trivedi, H.O. Sharma, L.S. Kothari, Phys. Rev. B **18**, 2668 (1978)
47. J.D. Filby, D.L. Martin, Proc. R. Soc. A **284**, 83 (1965)
48. C.A. Krier, R.S. Craig, W.E. Wallace, J. Phys. Chem. **61**, 522 (1957)

Quantum Groups, q-Oscillators and q-Deformed Quantum Mechanics

G. Vinod

Abstract The term *quantum group* was coined by the field medallist V. G. Drinfeld to describe a novel mathematical structure that made its first appearance in the quantum inverse scattering method. Later it found applications in diverse areas of mathematics and physics. In this article the mathematical structure of quantum group is described from the point of view of physics. q-deformed oscillators, which were introduced as the realisation of the quantum group $su_q(2)$, are discussed with emphasis on the algebra related to the quon algebra. A q-deformed quantum mechanics is developed and a q-Schrödinger equation is proposed. Physical implications of q-deformation are discussed and the applications of q-deformation to various fields are mentioned.

Keywords Inverse scattering · q-formed oscillator · Quantum groups · Lie algebra

1 Quantum Groups

1.1 Introduction

Quantum group is a mathematical structure which became a major research area in mathematics and theoretical physics in the last decade of the twentieth century. Quantum group made its first appearance in the physics literature in connection with the quantum inverse scattering method, a technique for studying integrable systems in quantum field theory and statistical mechanics. This structure made its appearance through the works of Kulish, Reshetikhin, Sklyanin Faddeev and Thakhatajan [1–4]. Faddeev observed that the Yang-Baxter Equation, which is a sufficient condition for solvability of two-dimensional Ising model of statistical mechanics is to quantum groups, just as the Jacobi identity is to classical Lie algebras. Drinfeld [4] realised that the algebraic structure associated with quantum inverse scattering method can

G. Vinod (✉)
Department of Physics, Sree Sankara College, Kalady, Ernakulam, Kerala, India
e-mail: vinodrohini@gmail.com

K. S. Sreelatha and V. Jacob (eds.), *Modern Perspectives in Theoretical Physics*,
https://doi.org/10.1007/978-981-15-9313-0_7

be reproduced by a suitable algebraic quantisation of Poisson Lie algebras. The same relations were obtained by Jimbo [5] through a somewhat different scheme. Drinfeld coined the term *quantum group* during the international Conference on Mathematics in Berkeley, in 1986. Drinfeld has been awarded the field medal in that ICM for his contributions to mathematics including quantum groups. Quantum groups or *quantum universal enveloping algebras* arise topologically in the theory of knot and link invariants [6] and geometrically in the study of non-commuting geometries [7].

1.2 Quantum Group as a Key to New Physics

From the viewpoint of physics, quantum group includes two basic ideas, namely, the deformation of an algebraic structure and the notion of a non-commutative co-multiplication [8]. The idea of deformation is familiar in physics: the Poincare group is a deformation of the Galilie group, which is regained in the limit $c \to \infty$. Also quantum mechanics can be considered as a deformation of classical mechanics which is regained in the limit $\hbar \to 0$. In the deformation of algebraic structure usually deformation parameter q is introduced and in the limit $q \to 1$, the original structure is regained. As a result of q-deformation, commutative algebra becomes a non-commuting one. This is the origin of the term *quantum* in quantum groups since quantization is in effect replacement of commuting things by non-commuting things.

The concept of co-multiplication is also inherent in quantum physics. For instance, consider the action of angular momentum operator **J** in quantum mechanics: Angular momentum is additive in both classical and quantum mechanics,

$$\mathbf{J}_{total} = \mathbf{J}_1 + \mathbf{J}_2 \tag{1}$$

In the ketspace formed from the product of the ket spaces spanned by the eigen kets of $\mathbf{J}_1$ and $\mathbf{J}_2$, the action of $\mathbf{J}_{total}$ can be expressed as

$$\mathbf{J}_{total} = \mathbf{J}_1 \otimes \mathbf{1} + \mathbf{1} \otimes \mathbf{J}_2 \tag{2}$$

This is actually a co-multiplication:

$$\Delta(\mathbf{J}) = \mathbf{J} \otimes \mathbf{1} + \mathbf{1} \otimes \mathbf{J} \tag{3}$$

Thus the vector addition of angular momentum in quantum mechanics defines a co-multiplication in a bialgebra. This is an example of commutative co-multiplication. The raising and lowering operators in angular momentum theory obey such a co-multiplication.

The z-component of total angular momentum J_0 and the ladder operators $J_\pm$ constitute an $su(2)$ algebra.

$$\begin{aligned} [J_0, J_\pm] &= \pm J_\pm \\ [J_+, J_-] &= 2J_0 \end{aligned} \tag{4}$$

A prototype q-deformation of this algebra is

$$\begin{aligned} [J_0, J_\pm] &= \pm J_\pm \\ [J_+, J_-] &= [2J_0]_q \end{aligned} \tag{5}$$

$[2J_0]_q$ is the q-deformation of $2J_0$, which in general depends on a parameter q. There are more than one definition of q-deformation in q-analysis all of which reproduces the original structure in the limit $q \to 1$. Thus different $su_q(2)$ are possible and all of them reduces to the $su(2)$ algebra in the limit $q \to 1$.

1.3 Formal Definition of a Quantum Group

Quantum groups are not groups; they are deformed Lie algebras. Formally, quantum groups are defined to be Hopf algebras, which are in general non-commutative [4–11] algebra is a bialgebra with an antipode. Bialgebra is a vector space which is an algebra as well as coalgebra. As an algebra is a way of multiplying things, a coalgebra is a way of unmultiplying things. Analogous to the notion of product in an algebra, there is the notion of co-product in a coalgebra. For a bialgebra $\mathcal{A}$ defined over a field k, product is defined as the mapping

$$m : \mathcal{A} \otimes \mathcal{A} \to \mathcal{A}$$

whereas co-product is defined by

$$\Delta : \mathcal{A} \to \mathcal{A} \otimes \mathcal{A}$$

unit is defined by

$$\eta : k \to \mathcal{A}$$

co-unit is defined by

$$c : \mathcal{A} \to k$$

The antipode ϵ is a linear map

$$\epsilon : \mathcal{A} \to \mathcal{A}$$

so that the following conditions are satisfied.

$$m\,(\epsilon \otimes id)\,\Delta = m\,(id \otimes \epsilon)\,\Delta = \eta \odot c$$

Quantum group is a bialgebra with an antipode, but with the conditions of commutativity and co-commutativity relaxed.

The concept of duality is a key point in defining the Lie algebra of a quantum group [12]. Given a Hopf algebra $\mathcal{A}(m, \Delta, \eta, c, \epsilon)$, the dual of $\mathcal{A}$, $\mathcal{A}^*$ will be endowed with the mappings m^*, Δ^*, η^*, c^*, ϵ^* such that

$$\begin{aligned}\langle m^*\,(f \times g)\,, x\rangle &= \langle (f \times g)\,, \Delta(x)\rangle \\ \langle \Delta^* f, x \times y\rangle &= \langle f, xy\rangle \\ \langle \eta^*\,(\alpha)\,, x\rangle &= \alpha\epsilon\,(x) \\ \epsilon^*\,(f) &= \langle f, 1\rangle \\ \langle S^* f, x\rangle &= \langle f, S(x)\rangle\end{aligned} \tag{6}$$

$\langle , \rangle$ denotes the pairing between a vector space and its dual.

There are mainly two different approaches to quantum groups:

- Lie algebraic approach developed by Drinfeld and Jumbo
- Non-commutative differential geometric approach developed by Manin.

1.4 Lie Algebraic Approach

Simple Lie algebras do not admit non-trivial deformations in the category of Lie algebras. Hence Drinfeld [4] and Jimbo [5] independently introduced the idea of deforming them in the category of Hopf algebras [10]. The resulting algebraic structure is called quantum universal enveloping algebra or popularly, quantum group, even though they are not at all groups.

There is no general prescription for defining the mappings for a given algebraic structure so as to make it a Hopf algebra. Consider the $sl(2)$ algebra formed by the generators X_+, X_- and H:

$$\begin{aligned}\left[H, X_\pm\right] &= \pm X_\pm \\ \left[X_+, X_-\right] &= 2H\end{aligned} \tag{7}$$

The co-product can be defined as

$$\begin{aligned}\Delta(H) &= H \otimes \mathbf{1} + \mathbf{1} \otimes H \\ \Delta(X_\pm) &= X_\pm \otimes \mathbf{1} + \mathbf{1} \otimes X_\pm\end{aligned} \tag{8}$$

This is a co-commutative co-product. Now assume that the deformed algebra has the form

$$\begin{aligned} \left[H, X_{\pm}\right] &= \pm X_{\pm} \\ \left[X_{+}, X_{-}\right] &= f(H) \end{aligned} \tag{9}$$

where $f(H)$ is arbitrary at this stage. Define co-multiplication in this algebra as

$$\begin{aligned} \Delta(H) &= H \otimes \mathbf{1} + \mathbf{1} \otimes H \\ \Delta(X_{\pm}) &= X_{\pm} \otimes f + g \otimes X_{\pm} \end{aligned} \tag{10}$$

Then the condition

$$(id \otimes \Delta)\, \Delta(X_{\pm}) = (\Delta \otimes id)\, \Delta(X_{\pm})$$

implies that

$$\begin{aligned} \Delta(f) &= f \otimes f \\ \Delta(g) &= g \otimes g \end{aligned} \tag{11}$$

These along with the definition of $\Delta(H)$ suggests the choice

$$f(H) = e^{\mu H}; \quad g(H) = e^{\nu H} \quad \mu, \nu \in \mathbb{R}$$

If we redefine $X_{\pm}$ by making an appropriate transformation, it is easy to show that

$$\begin{aligned} \Delta(X_{+}) &= X_{+} \otimes e^{\mu H} + \mathbf{1} \otimes X_{+} \\ \Delta(X_{-}) &= X_{-} \otimes \mathbf{1} + e^{-\mu H} \otimes X_{-} \end{aligned} \tag{12}$$

It can be shown that

$$\begin{aligned} \Delta\left[X_{+}, X_{-}\right] &= \left[\Delta X_{+}, \Delta X_{-}\right] \\ &= \left[X_{+}, X_{-}\right] \otimes e^{\mu H} + e^{-\mu H} \otimes \left[X_{+}, X_{-}\right] \end{aligned} \tag{13}$$

This relation along with the co-product of $f(H)$ suggests that $\left[X_{+}, X_{-}\right]$ may be deformed according to

$$\left[X_{+}, X_{-}\right] = \frac{e^{2\mu H} - e^{-2\mu H}}{e^{\mu H} - e^{-\mu H}} \tag{14}$$

If we put $e^{\mu H} = q$, the deformed algebra becomes

$$\begin{aligned} \left[H, X_{\pm}\right] &= \pm X_{\pm} \\ \left[X_{+}, X_{-}\right] &= \frac{q^{2H} - q^{-2H}}{q^{H} - q^{-H}} \end{aligned} \tag{15}$$

This is the standard form of $\mathcal{U}_q sl(2)$. Further if we assume that $H^\dagger = H$; $X_\pm^\dagger = X_\mp$, $\mathcal{U}_q sl(2) \to \mathcal{U}_q su(2)$. If we represent the generators by $J_0, J_+, J_-, \mathcal{U}_q su(2)$, or simply $SU_q(2)$ takes the form

$$\begin{aligned} \left[J_0, J_\pm\right] &= \pm J_\pm \\ \left[J_+, J_-\right] &= [2J_0]_q \end{aligned} \tag{16}$$

where

$$[N]_q = \frac{q^N - q^{-N}}{q - q^{-1}} \tag{17}$$

which approaches N in the limit $q \to 1$. There are two distinct expressions for the Casimir operator corresponding to integer and half-integer values of j in the representation of the algebra. They are $J_- J_+ + [J_0]_q [J_0 + 1]_q$ and $J_- J_+ + [J_0]_q [J_0 + \frac{1}{2}]_q$ respectively. Quantization of other Lie algebras have also appeared in the literature. Callegini et al. [12] have constructed $H_q(1)$ and $E_q(2)$ by contracting $SU_q(2)$. The existence of a Jacobi identity

$$[A, [B, C]] + [B, [C, A]] + [C, [A, B]] = 0 \tag{18}$$

is an essential requirement of any Lie algebra. In order to construct the q-analogue, Chaichian et al. [13] defined a q-deformed commutator

$$[A, B]_q = AB - qBA \tag{19}$$

They have observed that the following identity holds for arbitrary values of p and q:

$$\left[A, [B, C]_p\right]_q + q\left[B, [C, A]_p\right]_q + \left[C, [A, B]_{pq}\right] = 0 \tag{20}$$

1.5 Non-commutative Differential Geometric Approach

Manin [7] showed that a system of 2×2 matrices $T = \begin{pmatrix} a & b \\ c & d \end{pmatrix}$ with non-commuting matrix elements could be a quantum group provided the bialgebra $\mathcal{A}$ generated by a, b, c, d satisfy the following rules for multiplication:

$$\begin{aligned} ab &= qba; \quad ac = qca; \\ bd &= qdb; \, cd = qdc; \\ bc &= cb; \\ ad - da &= \left(q - q^{-1}\right) bc \end{aligned} \tag{21}$$

$$\Delta T = T \otimes T \tag{22}$$

If T_1 and T_2 are two 2×2 matrices with non-commuting elements, whose elements satisfy the relations (21), but the elements of T_1 commuting with those of T_2, then the elements of the matrix product $T_1 T_2$ also satisfy the relations (21). But conditions (21) can be obtained by proposing the idea of a quantum plane, in which the coordinates do not commute:

$$xy = qyx \tag{23}$$

If x and y commute with the matrix elements a, b, c, d satisfying the relations (21), then $x\prime$, $y\prime$ defined by

$$\begin{pmatrix} x\prime \\ y\prime \end{pmatrix} = \begin{pmatrix} a & b \\ c & d \end{pmatrix} \begin{pmatrix} x \\ y \end{pmatrix} \tag{24}$$

will satisfy

$$x\prime y\prime = qy\prime x\prime \tag{25}$$

In other words, if $\mathcal{A}$ is the algebra generated by x and y with relations (23), and $\mathcal{H}$ that generated by a, b, c, d with relations (21), then the map $\delta : \mathcal{A} \to \mathcal{H} \otimes \mathcal{A}$ defined by

$$\delta \begin{pmatrix} x \\ y \end{pmatrix} = \begin{pmatrix} a & b \\ c & d \end{pmatrix} \otimes \begin{pmatrix} x \\ y \end{pmatrix} \tag{26}$$

is a homomorphism. Thus the relations (21) constitute a sufficient condition on the elements of the matrix T for the action $\mathbf{x} \to \mathbf{Tx}$ on a column vector $\mathbf{x}$ to preserve the relations (23) between the components of $\mathbf{x}$. The same is true for a row vector $\tilde{\mathbf{x}}$ and the action $\tilde{\mathbf{x}} \to \tilde{\mathbf{xT}}$.

Conversely, (21) are the necessary conditions for the quantum plane condition to be preserved for both column and row vectors: i.e., if

$$xy = qyx \Rightarrow x\prime y\prime \tag{27}$$

then a, b, c, d satisfy the relations (21). Thus the relations (21) are consequences of the non-commutativity of space.

Relations (21) define the quantum general linear group $C_q GL(2)$. If we further assume

$$ad - qbc = 1 \tag{28}$$

we get the quantum unimodular group $C_q SL(2)$.

If we consider the quantization of $G = GL(2)$, taking the algebra of functions on G to be the algebra H generated by the non-commuting matrix elements a, b, c, d defined by (21). Further, if we define

$$\mathcal{E} = \frac{\partial}{\partial b}|_I$$
$$\mathcal{F} = \frac{\partial}{\partial c}|_I$$
$$\mathcal{I} = \frac{\partial}{\partial a} + \frac{\partial}{\partial d}|_I$$
$$\mathcal{F} = \frac{\partial}{\partial a} + \frac{\partial}{\partial d}|_I \quad (29)$$

where I denotes the identity matrix. These relations imply

$$[\mathcal{H}, \mathcal{E}] = 2\mathcal{E}$$
$$[\mathcal{H}, \mathcal{F}] = -2\mathcal{F}$$
$$[\mathcal{I}, \bullet] = 0 \quad (30)$$
$$\mathcal{E}\mathcal{F} - q\mathcal{F}\mathcal{E} = \frac{q^{2\mathcal{H}} - q^{-2\mathcal{H}}}{q - q^{-1}} \quad (31)$$

Thus the elements $\mathcal{E}, \mathcal{F}, \mathcal{H}, \mathcal{I}$ generates the quantum Lie algebra $U_q sl(2)$. Thus, non-commutativity of space leads to quantisation of the Lie algebra.

2 Elements of q-Analysis

Classical q-analysis has deep roots down to the beginning of the nineteenth century. In q-analysis, the q-deformation of a number is given by [14]

$$[n]_q = \frac{q^n - 1}{q - 1} \quad (32)$$

Thus, $[1]_q = 1, \quad [0]_q = 0$, independent of the value of q. Also, as $q \to 1$, $[n]_q \to n$. This definition of q-deformation lacks $q \leftrightarrow q^{-1}$ symmetry. The additive inverse of the q-integer is defined by

$$[n]_q + q^n[-n]_q = 0 \quad (33)$$

The q-factorial is given by

$$[n]_q! = [n]_q[n-1]_q[n-2]_q \ldots [2]_q[1]_q \quad (34)$$

The q-exponential function is defined as

$$exp_q X = \sum_{n=0}^{\infty} \frac{X^n}{[n]_q!} \quad (35)$$

For q-exponential functions,$exp_q X \cdot exp_q Y \neq exp_q (X+Y)$. The q-analogues of trigonometric functions are defined as

$$\sin_q x = \frac{1}{2i}\left\{exp_q(ix) - exp_q(-ix)\right\} \tag{36}$$

$$\cos_q x = \frac{1}{2}\left\{exp_q(ix) + exp_q(-ix)\right\} \tag{37}$$

The q-difference operator $\mathcal{D}_x$ is defined by [15]

$$\mathcal{D}_x f(x) = \frac{f(qx) - f(x)}{x(q-1)} \tag{38}$$

This operator is defined on a q-lattice in which the lattice points are in a geometric sequence. This q-difference operator does not possess $q \leftrightarrow q^{-1}$ symmetry. As $q \to 1, \mathcal{D}_x \to \frac{d}{dx}$, if it exists. It is evidently a linear operator. Also, it is easy to show that

$$\mathcal{D}_x(x^n) = [n]_q x^{n-1} \tag{39}$$

Hence we have the following results:

$$\mathcal{D}_x exp_q(x) = exp_q(x) \tag{40}$$

$$\mathcal{D}_x \sin_q(x) = \cos_q(x) \tag{41}$$

$$\mathcal{D}_x \cos_q(x) = -\sin_q(x) \tag{42}$$

Hence $\sin_q(x)$ and $\cos_q(x)$ are solutions of the q-difference equation

$$\left(\mathcal{D}_x^2 + k^2\right) f(x) = 0 \tag{43}$$

The q-analogues of the Leibinitz product rule and the quotient rule are found as

$$\begin{aligned}\mathcal{D}_x\left(u(x)v(x)\right) &= u(qx)\mathcal{D}_x v(x) + v(x)\mathcal{D}_x u(x) \\ &= u(x)\mathcal{D}_x v(x) + v(qx)\mathcal{D}_x u(x)\end{aligned} \tag{44}$$

$$\mathcal{D}_x\left(u(x)/v(x)\right) = \frac{v(x)\mathcal{D}_x u(x) - u(x)\mathcal{D}_x v(x)}{v(qx)v(x)} \tag{45}$$

q-integration is defined by

$$\int_a^b f(x)d(qx) = (1-q)\left(b\sum_{r=0}^{\infty} q^r f(q^r b) - a\sum_{r=0}^{\infty} q^r f(q^r a)\right) \tag{46}$$

Product rule for this q-integral is

$$\int g(x)\mathcal{D}_x f(x)d(qx) = f(x)g(x) - \int f(qx)\mathcal{D}_x g(x)d(qx) \tag{47}$$

It is helpful to introduce the dilation operator $\hat{Q}$:

$$\hat{Q}f(x) = f(qx) \tag{48}$$

$$\begin{aligned}\hat{Q}x^m = q^m x^m &= e^{m\ln q}x^m \\ &= \left(1 + \frac{m\ln q}{1!} + \frac{(m\ln q)^2}{2!} + \cdots\right)x^m \\ &= \left(1 + \frac{(\ln q)x\partial_x}{1!} + \frac{((\ln q)^2(x\partial_x)^2}{2!} + \cdots\right)x^m \\ &= e^{(\ln q)x\partial_x}x^m = q^{x\partial_x}x^m\end{aligned} \tag{49}$$

This can be generalised as

$$q^{\pm ax\partial_x}f(x) = f(q^{\pm a}x) \tag{50}$$

With the help of the dilation operator, the action of the q-difference operator can be written as

$$\mathcal{D}_x f(x) = \frac{q^{x\partial_x}f(x) - f(x)}{x(q-1)} \tag{51}$$

The following q-commutation relation holds:

$$[\mathcal{D}_x, x]_q = 1 \tag{52}$$

Also it can be shown that

$$\left[\mathcal{D}_x, q^{x\partial_x}\right]_q = 0 \tag{53}$$

and

$$\left[\mathcal{D}_x, q^{-x\partial_x}\right]_{q^{-1}} = 0 \tag{54}$$

Generalisation of the above results is

$$\left[\mathcal{D}_x, q^{ax\partial_x}\right]_{q^a} = 0; \quad a \in \mathbb{Q} \tag{55}$$

$$\left[\mathcal{D}_x^n, q^{ax\partial_x}\right]_{q^{na}} = 0; \quad n \in \mathbb{Z}, \quad a \in \mathbb{Q} \tag{56}$$

An immediate consequence of (56) is

$$\mathcal{D}_x^2 q^{-x\partial_x} = \left(q^{-\frac{1}{4}}\mathcal{D}_x q^{-\frac{1}{2}x\partial_x}\right)^2 \tag{57}$$

Alternate definitions of q-basic number and q-difference operator exist in the literature:

$$[\mathfrak{n}]_\mathfrak{q} = \frac{q^n - q^{-n}}{q - q^{-1}} \tag{58}$$

$$\mathfrak{D}_x = \frac{f(qx) - f(q^{-1}x)}{q - q^{-1}} \tag{59}$$

These definitions have $q \leftrightarrow q^{-1}$ symmetry. Also, the q-difference operators $\mathcal{D}_x(q)$ and $\mathfrak{D}_x(q^{1/2})$ are related by

$$\mathcal{D}_x(q) = q^{\frac{1}{2}x\partial_x}\mathfrak{D}_x(q^{1/2}). \tag{60}$$

3 q-Deformed Oscillators

The generators of $su(2)$ can be realised by the creation and annihilation operators a and $a^\dagger$ in the form

$$J_+ = -1/2a^2; \quad J_- = 1/2a^{\dagger 2}; \quad J_0 = \frac{1}{4}\left(aa^\dagger + a^\dagger a\right) \tag{61}$$

where the operators a and $a^\dagger$ obey the commutation rule

$$\left[a, a^\dagger\right] = 1 \tag{62}$$

The generators of $su_q(2)$ can be realised if we replace the commutation relation (62) by a q-deformed commutation relation (q-CR) retaining relations (61) with the difference that $J_\pm$ and J_0 are now generators of $su_q(2)$ and a and $a^\dagger$ are q-deformed. Biedenharn[16] and Macfarlane [17] independently found two q-CRs for q-deformed operators $\tilde{a}$ and $\tilde{a}^\dagger$. They are

$$\tilde{a}\tilde{a}^\dagger - q^{\frac{1}{2}}\tilde{a}^\dagger\tilde{a} = q^{-\tilde{N}/2} \tag{63}$$

and

$$\tilde{a}\tilde{a}^\dagger - q^{-1}\tilde{a}^\dagger\tilde{a} = q^{\tilde{N}} \tag{64}$$

where the operators $\tilde{N}$ satisfies the following commutation relations:

$$\left[\tilde{N}, \tilde{a}\right] = -\tilde{a}; \quad \left[\tilde{N}, \tilde{a}^{\dagger}\right] = \tilde{a}^{\dagger} \tag{65}$$

The commutation relations (63) and (64) are equivalent and they can be written alternatively as [18]

$$\tilde{a}\tilde{a}^{\dagger} - q\tilde{a}^{\dagger}\tilde{a} = q^{-\tilde{N}} \tag{66}$$

From (65),

$$\begin{aligned} \tilde{a}q^{\pm rN} &= q^{\pm r}q^{rN}\tilde{a} \\ \tilde{a}^{\dagger}q^{\pm rN} &= q^{\mp r}q^{rN}\tilde{a}^{\dagger} \end{aligned} \tag{67}$$

where r is a rational number. If we make the substitutions

$$\tilde{a} = q^{-\frac{1}{4}\tilde{N}}a; \quad \tilde{a}^{\dagger} = a^{\dagger}q^{-\frac{1}{4}\tilde{N}} \tag{68}$$

and use Eqs. (65), (63) can be written as

$$aa^{\dagger} - qa^{\dagger}a = 1 \tag{69}$$

whose multimode generalisation is the quon algebra proposed by Greenberg [19].

$$a_i a_j^{\dagger} - q a_j^{\dagger} a_i = \delta_{ij} \tag{70}$$

We can construct the representation of (70) in the Fock space spanned by the orthonormalised eigenstates $|n\rangle$ of N.

$$|n\rangle = \frac{a^{\dagger^n}}{\sqrt{[n]_q!}}; \quad N|n\rangle = n|n\rangle; \quad a|0\rangle = 0 \tag{71}$$

In view of (32), (70) can be had with the following relations:

$$aa^{\dagger} = [N+1]_q; \quad a^{\dagger}a = [N]_q \tag{72}$$

The q-commutator (70) can be written as [20]

$$\left[a, a^{\dagger}\right] = f(N) \tag{73}$$

with

$$\begin{aligned} f(N) &= q^{N} \quad for \quad q \neq 0 \\ f(N) &= \theta(1-N) \quad for \quad q = 0 \end{aligned} \tag{74}$$

where θ is the Heaviside step function defined by

$$\begin{aligned} \theta(x) &= 1 \quad for \quad x > 0 \\ &= 0 \quad for \quad x \leq 0 \end{aligned} \tag{75}$$

4 q-Deformed Quantum Mechanics

The q-deformed harmonic oscillator inspire the search for a q-deformed quantum mechanics which produce the results of standard quantum mechanics when the deformation parameter approaches a particular value. The q-deformed calculus, when applied to quantum mechanics with its fundamental postulates preserved, gives rise to q-quantum mechanics. Quantum mechanics is usually deformed in two different ways: either one may replace the canonical commutation relations by a q-commutation relation or may replace the momentum operator in the Schrödinger equation by a q-deformed one. The form of the q-momentum operator depends on the definition of the q-difference operator that we use. Wess and Zumino [21] have constructed a q-deformed momentum operator which is Hermitian when q is a root of unity. Vinod et al. has developed a q-deformed Schrödinger equation with a non-Hermitian momentum operator [22].

4.1 General Formulation of q-Quantum Mechanics

In the linear vector space $\mathcal{E}$, we define a norm mapping $\mathcal{E} \rightarrow \mathbb{R}^+ : v \rightarrow \|v\|$ such that

$$\|\alpha v\| = |\alpha| \, \|v\|; \quad \|v + u\| \leq \|v\| + \|u\|; \quad \|v\| = 0 \Rightarrow v = 0$$

$$u, v \in \mathcal{E}, \alpha \in \mathbb{C}$$

For square integrable function in the position space, the norm can be written as

$$\|v\| = \left[\int v^*(x) v(x) d(qx) \right]^{1/2} \tag{76}$$

The scalar product of two vectors is defined as

$$\mathcal{E} \times \mathcal{E} \rightarrow \mathbb{C} : (u, v) \rightarrow \langle u | v \rangle$$

such that

$$\langle v|v\rangle = \|v\|^2; \quad \langle v|u\rangle = \langle u|v\rangle^* \quad \langle v|\alpha u\rangle = \alpha \langle v|u\rangle \quad \langle \alpha v|u\rangle = \alpha^* \langle v|u\rangle$$
$$\langle u|v+w\rangle = \langle u|v\rangle + \langle u|w\rangle; \quad \langle u+v|w\rangle = \langle u|w\rangle + \langle v|w\rangle$$
$$\langle v|v\rangle = 0 \ \ iff \ \ v = 0 \quad (77)$$

For functions in position space, the scalar product takes the form

$$\langle u|v\rangle = \int u^*(x)v(x)d(qx) \quad (78)$$

In addition to the properties (77), this integral satisfies the fundamental theorem of q-integral calculus, namely,

$$\int_a^b [\mathcal{D}_x f(x)]\, d(qx) = f(b) - f(a) \quad (79)$$

Two vectors u and v are said to be q-orthogonal if $\langle u|v\rangle = 0$.

A linear vector space is said to be *complete* if there exists a set of vectors in it such that every vector v in the space can be expressed as a convergent sum of these vectors. A complete linear space, with a norm defined as above is called a q-Banach space. The linear space $\mathcal{E}\prime$ of the continuous linear functionals of the q-Banach space is called the q-dual space of $\mathcal{E}$.

Let $\alpha\,(\mathcal{E}, \mathcal{F})$ denotes the space of continuous linear mappings of the q-Banach space $\mathcal{E}$ into the q-Banach space $\mathcal{F}$. Then $A \in \alpha\,(\mathcal{E}, \mathcal{F})$ induces a mapping $A^\dagger$: $\mathcal{F}\prime \to \mathcal{E}\prime$ known as the q-adjoint operator. The q-adjoint operator is unique and is defined by

$$\int u^*(x)Av(x)d(qx) = \int \left[A^\dagger u(x)\right]^* v(x)d(qx) \quad (80)$$

If

$$\int u^*(x)Av(x)d(qx) = \int \left[Au(x)\right]^* v(x)d(qx) \quad (81)$$

that is, if $A^\dagger = A$, A is said to be q-Hermitian. It can be shown that the eigenvalues of a q-Hermitian operator can take real values only; also, the eigenvectors of a q-Hermitian operator belonging to different eigenvalues are q-orthogonal.

4.2 q-Deformation of the Schrödinger Equation

The q-deformed time-independent Schrödinger equation is

$$H_q\psi_q = \{\frac{\mathbf{p}^2}{2m} + \mathbf{V}(\mathbf{x})\}\psi_q = E_q\psi_q \quad (82)$$

where $\mathbf{p}$ is the q-deformed momentum operator. In the coordinate representation of standard quantum mechanics, $-i\hbar\frac{d}{dx}$ serves as the one dimensional momentum operator. So it is natural to include the q-difference operator in the expression for the q-deformed momentum operator. Depending on the choice of the q-difference operator, there are several possibilities for the q-Schrödinger equation [21]. If we choose the q-difference operator (51), the q-deformed momentum operator will have the form $-i\hbar\mathcal{D}_x$. Using (80) and (47),

$$\mathcal{D}_x\dagger = \mathcal{D}_x q^{-x\partial_x} \tag{83}$$

Thus $\mathcal{D}_x$ is not q-Hermitian. However, $\mathcal{D}_x\mathcal{D}_x^\dagger$ will be q-Hermitian. Hence we take

$$\mathbf{p}^2 = -\hbar^2\mathcal{D}_x^2 q^{-x\partial_x} \tag{84}$$

Hence the q-deformed Schrödinger equation becomes [22]

$$\left(-\frac{\hbar^2}{2m}\mathcal{D}_x^2 q^{-x\partial_x} + \mathbf{V(x)}\right)\psi_q = E_q\psi_q. \tag{85}$$

4.3 Physical Implications of q-Deformation

From (57), we get

$$-\hbar^2\mathcal{D}_x^2 q^{-x\partial_x} = \left(\pm i\hbar q^{-\frac{1}{4}}\mathcal{D}_x q^{-\frac{1}{2}x\partial_x}\right)^2 \tag{86}$$

which indicates that the q-deformed momentum operator has the form

$$\mathbf{p} = -i\hbar q^{-\frac{1}{4}}\mathcal{D}_x q^{-\frac{1}{2}x\partial_x} \tag{87}$$

The sign is so chosen that in the $q \to 1$ limit, the expression in the standard quantum mechanics is regained. Thus for $q \neq 1$, the momentum operator is not Hermitian. However, the square of the momentum operator as well as the Hamiltonian are q-Hermitian. In standard quantum mechanics, only Hermitian or skew Hermitian operators can yield a Hermitian operator on squaring. In fact it can be shown that all odd powers of the momentum operator are non-q-Hermitian and all even powers are q-Hermitian. Since the momentum is the generator of translation, translational invariance is absent in q-quantum mechanics.

5 Conclusion

Though the interest in quantum groups and q-deformation started as early as 1989, they still remain hot areas of research in mathematics and physics. Since two major theories of the last century, namely, relativity and quantum mechanics involved some type of deformation, we expect this new mathematical structure will also lead to some new physics. The idea of non-commuting space fascinates mathematicians as well as physicists. It is reasonable to assume that a quantum theory of gravity will be associated with some type of deformation of the space-time structure. This opens up possibilities of q-deformed theories of gravitation [23, 24]. Besides giving rise to new paradigms in statistical mechanics [25, 26] and quantum optics [27, 28], q-deformed oscillators find phenomenological applications in areas like molecular physics [23, 24], condensed matter physics [29, 30] and nuclear physics [31, 32]. A large collection of publications in these areas are available in the literature and for brevity, I cite only a few. The study of q-deformed quantum mechanics will enhance our knowledge of the standard quantum mechanics.

The purpose of this article is to introduce the reader to one of the fast developing fields in theoretical physics and to mention Prof. K Babu Joseph's contribution in this field. His contribution is mainly in the application of the ideas of quantum groups and q-deformations to physical problems. It includes the application of q-oscillator to quantum optics [33], formulation of q-deformed quantum mechanics [22] and phenomenological applications in condensed matter physics [34].

References

1. P. Kulish, Y. Reshetikhin, J. Sov. Math. **23**, 2435 (1983)
2. E.K. Sklyanin, Func. Anal. Appl. **16**, 263 (1982)
3. Y. Reshetikhin, L.A. Takhtajan, L.D. Faddeev, Leningrad Math. J. **1**, 193 (1990)
4. V.G. Drinfeld, *Proceedings of the International Congress of Mathematicians* (Berkeley, 1986)
5. M. Jimbo, Lett. Math. Phys. **10**(63), 247 (1986)
6. V.F.R. Jones, Bull. Am. Mathemat. Soc. **12**, 103 (1985)
7. Y.I. Mannin, Ann. Inst. Fourier **37**, 191 (1987)
8. L.C. Biedenharn, Int. J. Theoret. Phys. **32**, 1789 (1990)
9. S. Majid, Int. J. Modern Phys. A **5**, 1 (1990)
10. V. Chari, D. Takur, Current Sci. **59**, 1297 (1990)
11. T. Tjin, Int. J. Modern Phys. A **8**, 231 (1992)
12. E. Calegini, R. Giachetti, E. Sorace, M. Tarlini, J. Math. Phys. **31**, 11 (1990)
13. M. Chaichian, R.G. Felipe, C. Montonen, J. Phys. A Math. Gen. **26**, 4017 (1993)
14. E. Heine, J. Math. **34**, 285 (1846)
15. H. Exton, *q-Hypergeometric Functions and Applications* (Ellis Horwood, Chichester, 1983)
16. L.C. Biedenharn, J. Phys. A Math. Gen. **22**, L873 (1989)
17. M.J. Macfarlane, J. Phys. A Math. Gen. **22**, 4581 (1989)
18. M. Chaichan, P. Kulish, Phys. Lett. **234B**, 72 (1990)
19. O.W. Greenberg, Phys. Rev. Lett. **64**, 705 (1990)
20. S. Chaturvedi, A.K. Kapoor, R. Sandhya, V. Srinivasan, R. Simon, Phys. Rev. A **43**, 4555 (1991)

21. J. Wess, B. Zumino, Nuclear Phys. B (Proc. Suppl.) **18B**, 302 (1990)
22. G. Vinod, K.B. Joseph, K.M. Valsamma, Pramana. J. Phys. **45**, 311 (1995)
23. S. Ikhdair, Chem. Phys. **361**, 9 (2009)
24. D. Bonatsos, B.A. Kotsos, P.P. Roychev, P.A. Terziev, Int. J. Quant. Chem. **95**, 1 (2003)
25. J.W. Goodison, D.J. Toms, Phys. Lett. A **19538** (1994)
26. K.-M. Shen, B.-W. Zhang, E.-K. Wang, Phys. A **487C**, 215 (2017)
27. P. Shanta, S. Chaturvedi, V. Srinivasan, J. Modern Optics **39**(6), 1301 (1992)
28. C. Quesne, K.A. Renson, V.U. Ikachuk, Phys. Lett. A **322**, 402 (2004)
29. X.Y. Hou, X. Huang, Y. He, H. Guo, arXiv:1804.02376
30. M.B. Harouni, R. Roknizadeh, M.H. Naderi, J. Phys. B At. Mol. Opt. Phys. **42**, 95501 (2009)
31. K.D. Svivatcheya, C. Babri, A.I. Georgieva, J.P. Drayayer, Phys. Rev. Lett. **93**, 152501 (2004)
32. D. Bonatsos, B.A. Kotsos, P.P. Roychev, P.A. Terziev, Phys. Rev. C **66**, 054306 (2002)
33. G. Vinod, K.B. Joseph, V.C. Kuriakose, Pramana. J. Phys. **42**, 299 (1994)
34. K.K. Leelamma, V.C. Kuriakose, K.B. Joseph, Int. J. Modern Phys. B **7**(14), 2697 (1993)

Mandelbrot Set and Fermat's Last Theorem

P. B. Vinod Kumar

Abstract In this article a brief description of the work done with Prof. K. Babu Joseph and some related problems are described. The notion of irrational points in the Mandelbrot set ($\mathcal{M}_2$) is explained. Fermat's last Theorem (FLT) om Topological Fields is discussed. We ask the question: Given n ϵZ_+ Do there exist $(x, y, z)\epsilon$ $\mathcal{M}_2 \times \mathcal{M}_2 \times \mathcal{M}_2$ such that $x^n + y^n = z^n$? An attempt is made to solve this problem.

Keywords Mandelbrot set · Julia set · Fermat's theorem

1 Mandlebrot Set

In this section some results related to the Mandelbrot set are given. Carleson and Gamelin [3] gives a wonderful description of Julia set and the Mandelbrot set. It is interesting to study the continuity of the map $\Phi : C \to R$, defined as $\phi(c) = \dim_H(J_c)$. Curt Mcmullen has studied some properties of the map in [1]. But we show that if f_c has a Siegel disc at c then ϕ is discontinuous at c. $M = \{c\epsilon C | J_c \text{ is connected}\}$ is named as the Mandelbrot set. There are a lot of mysteries behind Mandelbrot set. We will define irrational points on M using the Hausdorff dimension. Mcmullen has computed Hausdorff dimension of J_c for different values of c in [1].

Curt Mcmullen in [1] has proved the following result.

Theorem 1.1 [1] *(i) For $|c|$ small (near 0) the map $\phi : c \to dim_H(J_c)$ is analytic and $dim_H(J_c) \to 1 + \frac{|c|^2}{4log2}$, as $|c| \to 0$. $dim_H(J_c) \to \frac{2log2}{log|c|}$, as $|c| \to \infty$.*

A natural question arises at this point is about the map $\phi : C \to R$, defined as $\phi(c) = \dim_H(J_c)$. We will prove the following result.

P. B. Vinod Kumar (✉)
Department of Mathematics, Rajagiri School of Engineering and Technology,
Kochi, Kerala, India
e-mail: vinod_kumar@rajagiritech.edu.in

K. S. Sreelatha and V. Jacob (eds.), *Modern Perspectives in Theoretical Physics*,
https://doi.org/10.1007/978-981-15-9313-0_8

Theorem 1.2 [4] *Let $c \in M$ be a parameter value for which f_c has a Siegel disc. Then the map ϕ is discontinuous at c.*

Proof Let z_0 be a Siegel periodic point of f_c and denote $\triangle$ the Siegel disc around z_0, p its period, and θ the rotation angle. Let $U(c)$ be an open set containing c. By the Implicit function theorem, there exists a holomorphic mapping $\zeta : U(c) \to C$ such that $\zeta(c) = z_0$ and $\zeta(c)$ is the fixed number $(f_c)^p$. The mapping $\chi : c \to ((f_c)^p)'(\zeta(c))$ is holomorphic, hence it is either constant or open. If it is constant, all quadratic polynomials have a Siegel disc (see [2] in which it has proved that a quadratic polynomial has almost one cycle of Siegel discs). This is not possible: for instance, $f_{1/4}$ has a parabolic fixed point and thus no other non-repelling cycles. Therefore, χ is open and in particular there is a sequence of parameters $c_n \to c$ such that $\zeta(c_n)$ has multiplier $e^{\frac{2\pi p_n}{q_n}}$. Since $\zeta(c_n)$ is parabolic, it lies in the Julia set of J_{c_n}. Hence, $\text{dist}_H(Jc, J) \geq \frac{\text{dist}(c, \rho\triangle)}{2}$ for large n. Hence the result.

Converse of above result is not true.

For example, $c = 0.25$ is parabolic i.e., f_c has a parabolic cycle ($z = 0.25$). So it is geometrically finite but not expanding. As c decreases from $\frac{1}{4}$ along the real axis the map f_c undergoes a sequence of period doubling bifurcation at parabolic points c_n converging to the Feigenbaum point $c_F \approx -1.40115$. The map f_c is expanding for all $c \epsilon (c_F, \frac{1}{4}]$. At $c = \frac{1}{4}$, ϕ is discontinuous but f_c has no Siegel disc there.

In [1], Mcmullen proved the following result.

Theorem 1.3 [1] *The function ϕ is continuous in $(c_F, \frac{1}{4}]$.*

2 Fermat's Last Theorem

2.1 Topological Fields

Let $(F, *, .)$ be a Field and T be a topology on F. Then $(F, *, ., T)$ is called a topological Field. There are topological fields which look like as same if we view from different angles, for example, R^n.

A topological space F is homogenous if given any $x, y \in F$, there exist $f : F \to F$, a homeomorphism on to F such that $f(x) = y$.

In $(F, *, ., T)$ neighbourhood of 0 is called additive nucleius.

We prove the following results in this section.

Theorem 2.1 [5] *(i) Every topological field is homogenous. (ii) Every topological field is Hausdorff.*

Proof (i) Let $x, y \in F$. Let $L_a(b) = a.b$. Then $L_{yx}^{-1}(x) = y$. So given$x, y \in F$, there exists $L_{yx}^{-1} : F \to F$, which is a homeomorphism.

(ii) Let $x, y \in F. \Rightarrow x^{-1}.y \neq 0 \Rightarrow X - \{x^{-1}.y\}$ is a neighbourhood of 0. $\Rightarrow$ There is a symmetric neighbourhood $V \subset X - x^{-1}.y \Rightarrow x.V$ and $y.V$ are nbds of x and $y. \Rightarrow x.\ V \cap y.V \neq \phi$. So there exist $v_1, v_2 \in V$ such tha $tv_1 v_2^{-1} \in V$, which is a contradiction. Hence F is Hausdorff.

2.2 *Finite Fields*

Denote $F[P^n]$, where P is a prime n is a positive integer $F[p^n]$ has P^n elements. For n=1, fields are of the form $F(p) = \{0, 1, 2, \ldots, p-1\}$ multiplication and addition are usual operations, except multiples of p should be left out of the set (modulo p).

$$F(5) = \{0, 1, 2, 3, 4\}$$

All of the multiplications in the example are mod 5 because $P = 5$ addition on $F[4]$ is addition of polynomials. $(a + bx) + (c + dx) = (a + c) + x(b + d)$. What is the additive identity?

$$(a + bx) + (c + dx) = (a + bx) = (a + c) + x(b + d)$$

$a + c = 0, c = 0$ $b + d = b, d = 0$. Therefore 0 is still the additive identity. $(a + bx)$ has the additive inverse $-a - bx$. Usual multiplication of polynomials except need to use $x^2 + x + 1 = 0$

$$(a + bx) * (c + dx) = ac = x(bc + ad) + x^2(bd)$$

This seems to be a problem because $F[4]$ only has linear polynomials but we ended up with a quadratic one. When constructing the elements of $F[2]$ remember had multiplication modulo $p = 2$, so $F[4]$ as the "constant" polynomials. $\Rightarrow$ Remember when making these tables that each element will only show up once in any column or row. This is because we want to show that each element has a multiplicative inverse and that there are no zero divisors

$$F[9] = F[3^2] = \{a + bx | a, b \epsilon F[9]\}$$

. As a vector space, to find the multiplication table we need a monic-quadratic that has no zeros in $F[9]$. A monic-quadratic will have a coefficient of 1 on the highest degree term.

Try $x^2 + 0x + 1$, $f(0) = 1$, $f(1) = 2$, $f(2) = 2$ (no zeros!).

Note: The constant term must to be non-zero, because otherwise 0 is a zero. Also note that we only need one polynomial that works. From the above explanation, it is clear that in $F(p)$, FLT is true only if $n = (p - 1)$ or $n = \frac{p-1}{2}$.

If we change the metric on R^2, its geometry is different and hence FLT is no more true. For example, if we consider the taxi cab metric d on R^2

$$d((x_1, y_1), (x_2, y_2)) = |x_1 - x_2| + |y_1 - y_2|,$$

$$[d((0,0),(1,0))]^2 + [d((0,0),(0,1))]^2 \neq [d((1,0),(0,1))]^2.$$

Geometric shapes do not remain as it is if we change the topology of R^n. We will discuss the geometry of elliptic curves in the next section .

2.3 Elliptic Curves

Elliptic curve is simply the locus of points in the $x - y$ plane that satisfies an algebraic equation of the form $Y^2 = (X - x^n)(Y - y^n)$ (with some additional minor technical conditions). This is deliberately vague as to what sort of values x and y represent. In the most elementary case, they are real numbers, in which case the elliptic curve is easily graphed in the usual Cartesian plane. But the theory is much richer when x and y may be any complex numbers ($\in C$). And for arithmetic purposes, x and y may lie in some other field, such as the rational numbers Q or a finite field F.

So an elliptic curve is an object that is easily definable with simple high school algebra. Its amazing fruitfulness as an object of investigation may well depend on this simplicity, which makes possible the study of a number of much more sophisticated mathematical objects that can be defined in terms of elliptic curves.

It is very natural to work with curves in the complex numbers, since C is the algebraic closure of the real numbers. That is, it is the smallest algebraically closed field that contains the roots of all possible polynomials with coefficients in R. Being algebraically closed means that C contains the roots of all polynomials with coefficients in C itself. It's natural to work with a curve in an algebraically closed field, since then the curve is as "full" as possible.

The case of elliptic curves in the complex numbers is especially interesting, not only because of the algebraic completeness of C, but also because of the rich analytic theory that exists for complex functions. In particular, the equation of an elliptic curve defines y as an "algebraic function" of x. For every algebraic function, it is possible to construct a specific surface such that the function is "single-valued" on the surface as a domain of definition. It turns out that an elliptic curve, defined as a locus of points, is also the Riemann surface associated with the algebraic function defined by the equation.

So an elliptic curve is a Riemann surface. In fact, it is of a special type: a compact Riemann surface of genus 1. And not only that, but the converse is also true: every compact Riemann surface of genus 1 is an elliptic curve. In other words, elliptic curves over the complex numbers represent exactly the "simplest" sorts of compact Riemann surfaces with non-zero genus. Topologically, the genus counts the number of "holes" in a surface. A surface with one hole is a torus.

This topological equivalence of an elliptic curve with a torus is actually given by an explicit mapping involving the Weierstrass ρ-function and its first derivative.

This mapping is, in effect, a parameterization of the elliptic curve by points in a "fundamental parallelogram" in the complex plane.

The topological space that results from identifying opposite sides of a period parallelogram is called a complex torus. The fundamental periods that define the parallelogram generate a *lattice* in C consisting of all sums of integral multiples of ω_1 and ω_2. If L denotes the lattice, then $L = Z\omega_1 \oplus Z\omega_2$. The complex torus can then be described as C/L. What this all means, therefore, is that a (complex) torus is the "natural" domain of definition of the function or any doubly periodic complex function.

The Fermat equation is the prime example. In general, an elliptic curve has the form $y^2 = Ax^2 + Bx^2 + Cx + D$, but for considering arithmetical questions, it is natural to restrict our attention to the case where A, B, C, D are all rational. This assumption will usually be in effect when we are considering properties of elliptic curves involving arithmetical questions (as opposed to their more general analytic properties). If all coefficients are rational, the elliptic curve is said to be defined over Q. The all-important Taniyama-Shimura conjecture concerns only elliptic curves defined over Q.

The fact that any elliptic curve (not necessarily defined over Q) has an abelian group structure means that we can learn a lot about it by studying various of its subgroups. For considering arithmetical (i.e., number theoretic) questions, we restrict our attention to curves defined over Q. In that case, there are several interesting subgroups we can consider.

The first is the group of all points on the curve E which have an order that divides m for some particular integer m. That is, m "times" such a point is the identity element. Such points are called "m-division points", and the subgroup they make up is denoted E[m]. The reason for the name is that any point in E[m] generates a cyclic subroup of E (and E[m]) whose order divides m. If the order is actually m, then the points in the cyclic group generated by the point divide E into m segments.

It isn't necessarily the case that the coordinates of a point in E[m] have integral or rational coordinates. However, the coordinates will be algebraic numbers (i.e., roots of an algebraic equation with coefficients in Q). It's relatively easy to show that as an abstract group $E[m]$ is just the direct sum of two cyclic groups of order m i.e., $Z/mZ \oplus Z/mZ$, so its order is m. We shall see later that its real interest lies in the fact that we can construct representations of other groups of transformations that act on $E[m]$. Such representations will consist of 2×2 matrices with integral entries i.e., elements of $GL_2(Z)$.

Another interesting subgroup of E is the set of all points whose coordinates are rational. Such points are said to be rational points. If the curve is defined over Q, then it is a simple fact that the set of all rational points (if there are any) is a subgroup.

Another interesting subgroup of E is the set of all points whose coordinates are rational. Such points are said to be rational points. If the curve is defined over Q, then it is a simple fact that the set of all rational points (if there are any) is a subgroup.

The definition of $L(E, s)$ will be made based on details about a series of other groups connected with E. These arise by considering E as an elliptic curve over the finite fields F_P. This is the same as taking the original equation and reducing the coefficients mod p. If the equation of E has rational but non-integral coefficients, we would need to assume none of their denominators are divisible by p, so we might as well assume all coefficients to be integral to begin with (since if the denominators are prime to p they have inverses mod p). Further, the definition of an elliptic curve requires that there are no repeated roots of the polynomial in x, and this may fail to be true when reducing mod p for some primes. Such primes are said to have "bad reduction". There will be only a finite number of these for any particular curve (they will divide the discriminant), but they have to be dealt with specially.

For any prime p where E has good reduction, we can consider the elliptic curve $E(F_P)$ over F_P. Since F_P is finite, there are only a finite number of points on E(F_P)), so it is a finite group. The order of this group, $(E(F_P))$, turns out to be a very important number.

There is a general approach in number theory of trying to deal with "global" problems, such investigating the structure of $E(Q)$, by looking at a closely related "local" problem mod p for all primes p. This is why we are interested in $E(F_P)$. In particular, if $E(F_P)$ is "large" for most p, we would expect E(Q) to be large too.

We will see that the numbers $(E(F_P))$ are studied by relating them to coefficients of the Dirichlet series of $L(E, s)$, the L-function of E.

The most important fact about the minimal discriminant is that the primes which divide it are precisely the ones at which the curve has bad reduction. In other words, except for those primes, the reduced curve is an elliptic curve over F_P.

There is still another invariant of an elliptic curve E, called its conductor, and often denoted simply by N. The exact definition is rather technical, but basically the conductor is, like the minimal discriminant, a product of primes at which the curve has bad reduction. Recall that E has bad reduction when it has a singularity modulo p. The type of singularity determines the power of p that occurs in the conductor. If the singularity is a "node", corresponding to a double root of the polynomial, the curve is said to have "multiplicative reduction" and p occurs to the first power in the conductor. If the singularity is a "cusp", corresponding to a triple root, E is said to have "additive reduction", and p occurs in the conductor with a power of 2 or more.

If the conductor of E is N, then it will turn out that N is the "level" of certain functions called modular forms (not yet defined) with which, according to the Taniyama-Shimura conjecture, E is intimately connected.

If N is square-free, then all cases of bad reduction are of the multiplicative type. An elliptic curve of this sort is called *semistable*. It is for elliptic curves of this sort that Wiles proved the Taniyama-Shimura conjecture.

Theorem 2.2 *Any finite field with characteristic p has p^n elements for some positive integer n.*

Proof: Let L be the finite field and K the prime subfield of L. The vector space of L over K is of some finite dimension, say n, and there exists a basis $P_1, P_2, \ldots, P_n$ of L over K. Since every element of L can be expressed uniquely as a linear combination of the p_i over K i.e., every a in L can be written as a = $\sum \beta_i P_i$, with β_i in K, and since K has p elements, L must have p_n elements.

This theorem, while it does restrict the size of a finite field, does not say that one will exist for a particular power of a prime, nor does it specify how many finite fields can exist of a particular order. The answers to these questions can be deduced from the following theorem.

Theorem 2.3 *Let L be a field with characteristic p and prime subfield K. Then L is the splitting field for $F(x) = x^{P^n} - x$ iff L has P^n elements.*

Proof Suppose that L is the splitting field for $F(x) = x^{P^n} - x$ over K. Since $(f(x), f'(x)) = 1$, the roots of $f(x)$ are distinct and so L has at least p^n elements. Consider the subset

$$E = \{P \in L | P^{p^n} = P\}$$

of L. Clearly E contains p^n elements since it consists of the roots of $f(x)$. Suppose that $p, \beta \in E$; then $(P\beta)^{p^n} = (P)^{p^n}(\beta)^{p^n} = P\beta$ and hence, $P\beta$ in E. Also,

$$(P+\beta)^{p^n} = \sum \binom{p^n}{i} P^i \beta^{p^n-i} = P^{p^n} + \beta^{p^n} = P + \beta$$

Since p | $C(p^n, i)$ for $0 < i < p^n$, and hence $(p + \beta)$ in E. The existence of additive and multiplicative inverses is easy to show, so E is a subfield of L and also a splitting field for $f(x)$. Thus by 2.1 $E = L$ and L contains p^n elements.

Suppose now that L contains p^n. The multiplicative group of L, which we will denote by L^*, forms a group of order $p^n - 1$ and hence the order of any element of L^* divides $p^n - 1$. Thus $P^{p^n} = P$ for all P in L^* and the relation is trivially true for $P = 0$. Thus $f(x)$ splits in L.

Theorem 2.4 *$GF(p^n)*$ is cyclic*

Proof The multiplicative group $GF(p^n)*$is, by definition, abelian and of order $p^n - 1$. If $p^n - 1 = P_1^{e_1}, \ldots, p_k^{e_k}$, then, factoring $GF(p^n)*$ into a direct product of its Sylow subgroups, we have $GF(p^n)* = S(p_1) \times \ldots \times S(p_k)$, where $S(p_i)$ is the Sylow subgroup of order $(p_i)^{e_i}$. The order of every element in $S(p_i)$ is a power of p_i and let a_i in $S(p_i)$ have the maximal order, say$(p_i)_i^{e'}$, $e_i' \leq e_i$, for $i = 1, \ldots, k$. Since $(p_i, p_j) = 1$, i not equal j, the element $a = a_1 a_2 \ldots a_k$ has maximal order $m = (p_1)^{e'-1} \ldots (p_k)_k^{e'}$ in $GF(p^n)*$. Furthermore, every element of $GF(p^n)*$ satisfies the polynomial $x^m - 1$, implying that $m \geq p^n - 1$. Since $a \in GF(p^n)*$ has order m, m divides $p^n - 1$ and so, $m = p^n - 1$. Thus the element a is a generator and $GF(p^n)^*$ is cyclic.

A generator of $GF(p^n)*$ is called a *primitive* element of $GF(p^n)$.

The following theorem has some useful consequences.

Theorem 2.5 *Over any field K, $(x^m - 1) \mid (x^n - 1)$ iff m divides n*

Proof If $n = qm + r$, with $r < m$, then by direct computation

$$x^n - 1x^r \left(\sum_{i=0}^{q-1} \right) (x^m - 1) + (x^r - 1)$$

It follows that $(x^m - 1) \mid (x^r - 1)$ iff $x^r - 1 = 0$. i.e., $r = 0$

Theorem 2.6 *$GF(p^m)$ is a subfield of $GF(p^n)$ iff m divides n.*

Proof Suppose $GF(p^m)$ is a subfield of $GF(p^n)$; then $GF(p^n)$ may be interpreted as a vector space over $GF(p^m)$ with dimension, say, k. Hence, $p^n = p^{km}$ and $m|n$.

Now suppose $m|n$, which from the previous theorem and its corollary implies that $(x^{p^m-1} - 1) \mid (x^{p^n-1} - 1)$. Thus every zero of $x^{p^m} - x$ that is in $GF(p^m)$ is also a zero of $x^{p^n} - x$ and hence in $GF(p^n)$. It follows that $GF(p^m)$ is contained in $GF(p^n)$. Notice that there is precisely one subfield of $GF(p^n)$ of order p^m, otherwise $x^{p^m} - x$ would have more than p^m roots.

Although we will not prove it, the automorphism group of a finite field is cyclic. The standard generator of this group is the so-called *Frobenius automorphism* defined for a finite field of characteristic p as the map $x \to x^p$ for all x in $GF(p^n)$.

3 FLT on the Mandelbrot Set

Analogus to FLT we try to get three points x, y, z in the mandelbrot set so $x^n + y^n = z^n$; for all $n \geq 2$. If n is small, there are large number of points. As n increases x^n goes out side the mandelbrot set for $x \in \mathcal{M}_2$.

More concepts on periodic bulbs and external rays can be seen in [7].

The following result is taken from [8].

Theorem 3.1 *[8] There exists solution for the equation $x^n + y^n = z^n$ in M_2 for every $n \epsilon Z_+$ iff $\{x, y, z\} \in n/2n + 1$ bulb of M_2.*

Acknowledgements I thank Dr. Thirvikraman and Dr. K. Babu Joseph who participated in the discussion of this paper. I sincerely acknowledge the support given by the management of Rajagiri School of Engineering and Technology, India.

References

1. C. Mcmullen, Am. J. Math. **120**, 691–721 (1998)
2. A. Douady, A.M.S. Proc, Symp. Appl. Math. **49**, (1994)
3. L. Carleson, T. Gamelin, *Complex Dynamics* (Springer, 1993)
4. P.B. Vinod Kumar, K. Babu Joseph, Fractals **13**(3), 233–236 (2005)
5. P.B. Vinod Kumar, K. Babu Joseph, *Fermat's Last Theorem on Topological Fields*. https://arxiv.org/ftp/arxiv/papers/0802/0802.2439.pdf
6. A. Korgzik, *Fermat's Last Theorem* (Springer)
7. R.L. Devaney, Illuminating the Mandelbrot set, 13. https://math.bu.edu/people/bob/papers/mar_athan.pdf
8. P.B. Vinod Kumar, Solving $x^n + y^n = z^n$ inside the Mandelbrot set. (communicated)

Nonlinear Dynamics

Soliton Propagation Through Photorefractive Media

Lakshmi Parameswar

Abstract Incoherent soliton-beam propagation in photorefractive materials is investigated using the Manakov model. The Coupled Nonlinear Schrodinger Equation (CNLSE) is solved to find the conditions under which the soliton-like waves can propagate and the coupled-beam propagation through this media is investigated using two-beam coupling. With appropriate choice of experimental parameters, the input beam profile can be made to converge asymptotically to a soliton state whose PR nonlinearity compensate for diffraction, and beam profile remains unchanged as it propagates.

Keywords NLSE · Optical solitons · Photorefractive media · Two-beam coupling

1 Introduction

The development of mathematical properties of a large class of solvable nonlinear evolution equations has started with the first recorded observation of the "great solitary" wave by Scott Russell in 1834. The other class of nonlinear evolution equations includes the Korteweg de-Vries (KdV), the sine-Gordan (sG), and the Nonlinear Schrödinger Equations (NLSE) [1]. Nonlinear differential equations fall into two broad categories, the so-called "integrable" and "nonintegrable" systems. As a consequence, in nonlinear dynamics there emerged two prominent sub-disciplines "Solitons" and "Chaos", corresponding to the integrable and nonintegrable types, respectively. The "solitonic systems" are characterized by regular or predictable behavior for all times of evolution. Contrary to this "chaotic systems" exhibit completely irregular and unpredictable behavior with sensitive dependence on initial conditions.

L. Parameswar (✉)
Department of Physics, D. B. Pampa College, Parumala, Pathanamthitta, Kerala, India
e-mail: lakshmisuresh48@gmail.com

K. S. Sreelatha and V. Jacob (eds.), *Modern Perspectives in Theoretical Physics*,
https://doi.org/10.1007/978-981-15-9313-0_9

2 The Soliton and Its History

In nature not all waves disperse or spread and hence diminish over distances. But there are many cases of fairly permanent and powerful waves that have the capacity to travel extraordinary distances without diminishing in size or shape, and which are described by nonlinear equations of dispersive type. This kind of a wave was first observed by the Scottish naval architect "Scott Russel" in 1834. In 1830s Scott Russel carried out investigations on the shape of hulls of ships and observed the speed and forces needed to propel them. In 1834, riding on a horseback, he observed the "Great wave of translation" in the union canal and reported his observations to the British Association [2].

> I believe I shall best introduce the phenomenon by describing the circumstances of my own acquaintance with it. I was observing the motion of a boat which was rapidly drawn along a narrow channel by a pair of horses, when the boat suddenly stopped – not so the mass of water in the channel which it had put in motion, it accumulated round the prow of the vessel in a state of violent agitation, then suddenly leaving it behind, rolled forward with great velocity, assuming the form of a large solitary elevation, a rounded, smooth and well-defined heap of water, which continued its course along the channel apparently without change of form or diminution of speed. I followed it on horseback, and overtook it still rolling on at a rate of some eight or nine miles an hour, preserving its original figure some thirty feet long and a foot to a foot and a half in height. Its height gradually diminished, and after a chase of one or two miles I lost it in the windings of the channel.

This rolling pile of water is a solitary wave which maintained its shape and speed much larger than the convectional wave. Scott Russel also performed laboratory experiments [2] generating solitary waves by dropping a weight at one end of the water channel and obtained the relation, $c^2 = g(h + a)$ where h is the undisturbed depth of water, a is the amplitude of the wave, g the acceleration due to gravity and c the speed of the wave.

A simple nonlinear dispersive wave equation whose modern version is given by

$$U_t + 6\, U U_{xx} + U_{xxx} = 0 \tag{1}$$

where $(U U_{xx})$ is then nonlinear term and (U_{xxx}) is the dispersion term.

The solution is written as

$$U(x, t) = \frac{c}{2} \text{sech}^2 (\frac{\sqrt(c)(x - ct)}{2}) \tag{2}$$

A delicate balance between the nonlinear term $(U U_{xx})$ and dispersion term in equation [3] results in a solitary wave pulse which moves with uniform velocity proportional to the amplitude. Another important observation is that after interaction, such solitary waves emerge unaffected in their amplitude and velocity, except for phase shifts. A solitary wave can arise when the nonlinearity balances linear dispersion. In appropriate nonlinear systems, these solitary waves can interact elastically like particles without changing their shape and velocities.

3 Optical Solitons

Nonlinear optics is the study of phenomena that occur as a consequence of the modification of the optical properties of a material system by the presence of light. The beginning of the field of nonlinear optics is often taken to be the discovery of second harmonic generation by Franken et al. in 1961, shortly after the demonstration of the first working of laser by Maiman in 1960 [4]. Nonlinear optical phenomena are nonlinear in the sense that they occur when the response of a material system to an applied optical field depends in a nonlinear manner upon the strength of the optical field.

After the invention of lasers, nonlinear optics has emerged as one of the most sought-after subjects in all the frontiers of science by both theoreticians and experimentalists. Nonlinear optics has stirred many phenomena like fabrication of new nonlinear materials, harmonic generations, optical solitons, parametric amplification, stimulated Raman scattering, self-induced transparency, modulation instability, etc. which find a myriad of applications ranging from high data transmission in optical communication, switching amplifiers, pulse reshaping, pulse compression, tunable lasers to encoded message transmission. Notable among them is the concept of *optical soliton* pioneered by Hasegawa of Japan [5]. It revolutionalized the scope of telecommunication world and solitons are nowdays perceived to be carriers of communication signals in near future. Optical solitons were first observed by Mollenaeur and his group in 1980s in optical fibers [6].

Optical solitons fall into two categories (1) Spatial optical solitons and (2) Temporal optical solitons [7]. Spatial solitons are optical beams that can propagate in a nonlinear medium without diffraction, that is, their beam diameter remains invariant during propagation. A spatial soliton represents an exact balance between diffraction and nonlinearity induced self lensing or self defocusing effects. Spatial solitons have become one of the research areas in optics and nonlinear science. A temporal soliton is formed when group velocity dispersion (GVD) is totally counteracted by temporal self-focusing or self-phase modulation (SPM) effects [4, 5]. Optical temporal solitons have become a candidate for optical communication networks.

All optical solitons require a strong enough nonlinear interaction between themselves and the material in which they propagate. This interaction requires the so-called diffraction length for the spatial case or dispersion length for temporal (fiber) case which is comparable to a nonlinear length that characterizes self-focusing in the medium. In fibers low losses allow propagation distance of kilometers, and as a result, the very weak glass nonlinearity becomes cumulatively sufficient for soliton formation. In spatial case, however, the sample sizes are typically limited to centimeters, and thus, either the nonlinearities or the operator powers need to be larger. Therefore dimensionality is the factor that counterparts spatial solitons from fiber. Fiber solitons are described by (1+1) dimensional creatures. The higher dimensionalities of spatial solitons lead to host of interesting phenomena and processes, which have no analogue whatsoever in temporal case. These include, for example, full three-dimensional (3D) interaction between solitons and soliton spiraling, vortex solitons, angular momentum effects, rotating dipole vector solitons, etc.

4 Nonlinear Schrödinger Equation (NLSE) and Optical Solitons

The study of optical wave propagation in a nonlinear dispersive (dielectric) has been receiving considerable attention in recent times as the fiber can support under suitable circumstances a stable pulse called optical soliton [2–8] which arises due to compensation of the effect of dispersion of the pulses by nonlinear response of the medium. The analysis of such pulse propagation starts from the Maxwell's equation for the electromagnetic wave propagation in a dielectric medium given by where $\tilde{P}$ is the induced polarization given in (3), E represents the electric field, c is the velocity of light, μ_0 is the permeability of free space.

$$\nabla^2 E - \frac{1}{c^2}\frac{\partial^2 E}{\partial t^2} = -\mu_0 \frac{\partial^2 \tilde{P}}{\partial t^2} \tag{3}$$

In order to analyze Eq. (3), the following assumptions are taken into account. (1). The nonlinear part of induced polarization is treated as a small perturbation to the linear part. (2) The optical field is assumed to maintain its polarizability along the fiber. (3) Fiber loss is assumed to be small. (4) The nonlinear response of the fiber is assumed to be instantaneous. (5) In a slowly varying envelope approximation for the pulse propagation along fiber, the electric field can be written as

$$E(r,t) = \frac{1}{2}\hat{e}[F(x,y)E(z,t)e^{tk_0 z - \omega_0 t} + c.c] \tag{4}$$

where $\hat{e}$ is the unit polarization vector of light to be linearly polarized, $E(r,t)$ is the slowly varying electric field, $F(x,y)$ is the mode distribution function in the (x,y) plane, while k_0 and ω_0 denote the propagation constant and central frequency of the optical pulse.

Rewriting Maxwell's (13) by using the method of separation of variables and introducing the co-ordinate system, $T = t - \frac{z}{V_g}$, moving with the pulse at the group velocity $V_g = \frac{\partial k}{\partial \omega}$, the wave equation for the evolution of E is obtained as

$$i\frac{\partial E}{\partial k} - \frac{k''}{2}\frac{\partial^2 E}{\partial T^2} + \gamma_0 |E|^2 E = 0 \tag{5}$$

where $\gamma_0 = \frac{n_2 \omega_0}{cA_{eff}}$ Here A_{eff} denotes the effective core area of the single mode fiber, n_2 represents nonlinear refractive index coefficient. The parameter $k'' = \frac{\partial^2 k}{\partial \omega_0^2} = -\frac{1}{V_g^2}(\frac{\partial V_g}{\partial \omega})$ (at $\omega = \omega_0$) accounts for GVD. After normalizing Eq. (5) and using the transformations

$$q = \left(\frac{\gamma_0 T_0^{2\frac{1}{2}}}{|k^*|} E \right)$$
$$\xi = \frac{z|k''|}{T_0^2} \tag{6}$$
$$\tau = \frac{T}{T_0}$$

Redefining ξ as z and τ as t, we get the NLSE,

$$iq_x - sgn(k'')q_{tt} + 2|q^2|q = 0 \tag{7}$$

where T_0 represents the width of the incident pulse, z and t are the normalized distance and time along the direction of propagation and q, the normalized envelope. Interchanging t and x, the standard form of NLSE is obtained as

$$iq_t + q_{xx} \pm 2|q|^2 q = 0 \quad (q \in c) \tag{8}$$

If we consider normal dispersion regime [negative sign in (11)], the solution take "tanh" form and the intensity profile associated with such solutions shows dip in a uniform background and thus the solutions are called "dark solitons" [9].

5 Coupled Nonlinear Schrödinger Equation (CNLSE)

The formation of optical soliton is due to interplay between spreading of pulse and nonlinear response of medium [Kerr effect] which leads to an intensity dependent phase change described by self-phase modulation [SPM] of incident pulse. In the case of birefringent fibers in addition to SPM one has to consider cross phase modulation [XPM] which leads to a phase dependence of each mode on the intensity of co-propagating modes. The resulting propagation equation is a set of coupled nonlinear Schrödinger equations [CNLSE] [10]. This equation was developed by S. V. Manakov in 1973 known as the Manakov Model. This equation describes two-mode propagation in optical fiber.

$$iq_{1x} + c_1 q_{1tt} + 2(\alpha|q_1|^2 + \beta|q_1|^2)q_2 = 0 \tag{9}$$
$$iq_{1x} + c_1 q_{1tt} + 2(\alpha|q_1|^2 + \beta|q_2|^2)q_1 = 0$$

c_1, c_2, α, β real parameters.

For the two specific parametric choices, Manakov system is defined by

$$iq_{2x} + q_{2tt} + 2(\alpha|q_1|^2 + \beta|q_1|^2)q_2 = 0 \quad (10)$$
$$iq_{1x} + q_{1tt} + 2(\alpha|q_1|^2 + \beta|q_2|^2)q_1 = 0$$

This model is not only useful to describe soliton propagation in optical fibers, but can also be used to describe incoherent soliton-beam propagation in the so-called photorefractive materials. These materials can often exhibit high nonlinearity with very low intensity optical pulses, even with milliWatt power lasers or incandescent bulbs. Examples of such photorefractive materials include lithium niobate ($LiNbO_3$) and strontium barium niobate ($SrBaNbO_3$) crystals. The coupled-beam propagation through photorefractive media is investigated using the theory of two-beam coupling [3] and the possibility of formation of dark as well as bright soliton is studied subject to some parametric conditions.

6 Photorefractive Solitons

Segev et al. [9] have shown that the photorefractive media can support a new type of soliton in which the mixing of two waves or self scattering is balanced by diffraction. The photorefractive (PR) soliton possess the property of independence absolute high intensity and can experience, absorption with no change in the transverse structure. The shape of the soliton modulates the refractive index via photorefractive (PR) effect, which results in exact compensation for the effect of diffraction and causes the light beam to propagate with an unvarying profile. The PR solitons have become a subject of increased interest for the past few years [9–13]. Chen et al. [14] showed that two mutually incoherent beams of same polarization can form coupled steady-state spatial solitons pair in a biased photorefractive medium. In general a coupled PR soliton pair is analogous to a Manakov soliton, in which both beams are mutually trapped [15]. It was also shown that a system of incoherently coupled equations may possess general solutions for coupled solitary waves. The condition in which two beams of light of slightly different frequencies interact in such a manner that energy is transferred from one beam to another is known as "two-beam coupling". CNLSE is used to describe the possibility of soliton-like propagation in PR media. The method used is two-beam coupling. Starting from the wave equation, the CNLSE is derived. We found that under certain conditions, this equation possesses soliton solutions.

7 Two-Beam Coupling Using Solitons

Consider two beams of light (which in general have different frequencies) interacting in a photorefractive material. To describe this process mathematically, the total electric field within the nonlinear medium is described as [11]

$$\tilde{E}(z,t) = A_1(z,t)\, e^{i(k_1.z-\omega_1 t)} + A_2(z,t)\, e^{i(k_2.z-\omega_2 t)} + c.c \tag{11}$$

$k_i = \frac{n_0\omega_i}{c}$, n_0 denoting the linear part of the refractive index experienced by each wave. The intensity distribution associated with the interference between the two waves is given by

$$I = \frac{n_0 c}{4\pi}\overline{\tilde{E}}^2 \tag{12}$$

where the over bar denotes an average over a time interval of many optical periods. The intensity distribution for $\tilde{E}(z,t)$ given by Eq. (11) is given by

$$I = \frac{n_0 c}{2\pi}\left\{A_1 A_1^* + A_2 A_2^* + \left[A_1 A_2^* e^{i(q.z-\delta t)}\right]\right\} \tag{13}$$

where we have introduced the wave vector difference

$$q = k_1 - k_2 \tag{14}$$

and the frequency difference

$$\delta = \omega_1 - \omega_2 \tag{15}$$

For the geometry we have assumed that $|\delta| << \omega_1$. The pattern moves upward for $\delta < 0$, moves downward for $\delta > 0$, and is stationary for $\delta = 0$. For a nonlinear medium, a refractive index variation is accompanied by this intensity variation. To allow the possibility of energy transfer, assume that the nonlinear part of the refractive index (n_{NL}) obeys a Debye relaxation equation of the form

$$\tau\frac{dn_{NL.}}{dt} n_{NL} = n_2 I \tag{16}$$

Solving Eq. (16) by the Green's function method,

$$n_{NL} = \frac{n_2}{\tau}\int_{-\infty}^{t} I(t')e^{t-t'}dt' \tag{17}$$

Introducing Eq. (13) for intensity into this equation, it is found that I varies as $e^{-i\delta t}$ leading to an integral of the form

$$\int_{-\infty}^{t} e^{-i\delta t'} e^{\frac{t'-t}{\tau}} dt' = \int_{-\infty}^{t} e^{(-i\delta+\frac{1}{\tau})t'} dt' = \frac{e^{-i\delta t}}{-i\delta+\frac{1}{\tau}} \tag{18}$$

Equation (17) hence shows that the nonlinear contribution to the refractive index is given by

$$n_{NL} = \frac{n_0 n_2 c}{2\pi}\left[(A_1 A_1^* + A_2 A_2^*) + \frac{A_1 A_2^* e^{i(q.r-\delta t)}}{1-i\delta t} + \frac{A_1^* A_2 e^{i(q.r-\delta t)}}{1+i\delta t}\right] \tag{19}$$

Like all electromagnetic phenomena, the propagation of optical fields through a nonlinear photorefractive medium is governed by Maxwell's equations. These equations are

$$\nabla.\tilde{D} = 4\pi\rho \tag{20}$$

$$\nabla.\tilde{B} = 0 \tag{21}$$

$$\nabla \times \tilde{E} = -\frac{1}{c}\frac{\partial \tilde{B}}{\partial t} \tag{22}$$

$$\nabla \times \tilde{H} = \frac{1}{c}\frac{\partial \tilde{B}}{\partial t} + \frac{4\pi}{c}\tilde{J} \tag{23}$$

Putting

$$\tilde{\rho} = 0, \tilde{J} = 0, \tilde{B} = \tilde{H} \tag{24}$$

Using Eqs. (24), (23) becomes

$$\nabla \times \tilde{B} = \frac{1}{c}\frac{\partial \tilde{D}}{\partial t} \tag{25}$$

Taking curl of Eq. (22) and using Eq. (25), the optical wave equation obtained is

$$\nabla^2 \tilde{E} - \frac{1}{c^2}\frac{\partial^2 \tilde{D}}{\partial t^2} = 0 \tag{26}$$

Introducing Fourier Transforms

$$\tilde{E}(z,t) = \int_{-\infty}^{\infty} E(z,t) e^{-i\omega t}\frac{d\omega}{2\pi} \tag{27}$$

and

$$\tilde{D}(z,t) = \int_{-\infty}^{\infty} D(z,t) e^{-i\omega t}\frac{d\omega}{2\pi} \tag{28}$$

where the Fourier amplitudes are related by

$$D(z,\omega) = \epsilon(\omega)E(z,\omega) \tag{29}$$

and $\epsilon(\omega)$ is the effective dielectric constant that describes both the linear and nonlinear contributions to this response. Using Eqs. (27), (28) and (29) in (26), the equation obtained is

$$\frac{\partial^2 E(z,\omega)}{\partial z^2} + \epsilon(\omega)\frac{\omega^2}{c^2}E(z,\omega) = 0 \tag{30}$$

Writing the Fourier transform of $A(z,t)$as

$$A(z,\omega') = \int_{-\infty}^{\infty} A(z,t)e^{-i\omega t}dt \tag{31}$$

which is related to $E(z,\omega)$ by

$$E(z,\omega) \approx A(z,\omega-\omega_0)e^{ik_0 z} \tag{32}$$

where the second, approximate form is obtained by noting that a quantity such as $\tilde{A}(z,t)$ which varies slowly in time cannot possess high frequency Fourier components. This expression for $E(z,\omega)$is introduced into Eq. (30) and slowly varying approximation is made, so that the term containing $\frac{\partial^2 A}{\partial z^2}$ can be dropped. The equation obtained for the signal and pump beams is

$$2ik_1\frac{\partial A_1}{\partial z} + (k^2 - k_1^2)A_1 = 0 \tag{33}$$

$$2ik_2\frac{\partial A_2}{\partial z} + (k^2 - k_2^2)A_2 = 0 \tag{34}$$

Generally

$$2ik_0\frac{\partial A}{\partial z} + (k^2 - k_0^2)A = 0 \tag{35}$$

where

$$k(\omega) = \sqrt{\epsilon(\omega)}\frac{\omega}{c} \tag{36}$$

In practice, k typically differs from k_0 by only a small fraction amount, and thus to a good approximation $k^2 - k_0^2$ can be replaced by $2k_0(k - k_0)$, so that Eq. (35) becomes

$$\frac{\partial A(z,\omega-\omega_0)}{\partial z} - i(k - k_0)A(z,\omega-\omega_0) = 0 \tag{37}$$

The propagation constant k depends on both frequency and intensity of optical wave. This dependence in terms of truncated power series can be written as

$$k = k_0 + \Delta k_{NL} + k_1(\omega-\omega_0) + \frac{1}{2}k_2(\omega-\omega_0)^2 \tag{38}$$

Here, we have introduced the nonlinear contribution to the propagation constant, given by

$$\Delta k_{NL} = \frac{\Delta n_{NL}\omega_0}{c} = \frac{n_2 I\omega_0}{c} \tag{39}$$

with

$$I = \frac{n_{lin}\omega_0 c}{2\pi}|A(z,t)|^2 \tag{40}$$

and

$$k_1 = \frac{dk}{d\omega} = \frac{1}{c}\left[n_{lin}(\omega) + \omega\frac{dn_{lin}}{d\omega}\right]_{\omega=\omega_0} = \frac{1}{v_g\omega_0} \tag{41}$$

and

$$k_2 = \frac{d^2k}{d\omega^2} = \frac{d}{d\omega}\left[\frac{1}{v_g(\omega)}\right] = \left(-\frac{1}{v_g^2}\frac{dv_g}{d\omega}\right)_{\omega=\omega_0} \tag{42}$$

In the above expression, k_1 is the reciprocal of the group velocity, and k_2 is a measure of the dispersion of the group velocity. Equation (37) for is next introduced into the reduced wave Eq. (36) and the equation becomes

$$\frac{\partial A}{\partial z} - i\Delta k_{NL}A - ik_1(\omega-\omega_0)A - \frac{1}{2}ik_2(\omega-\omega_0)^2A = 0 \tag{43}$$

which is transformed from the frequency domain to the time domain. For this multiply each term by the factor $e^{-i(\omega-\omega_0)t}$ and integrate the resulting equation over all values of $(\omega-\omega_0)$. The resulting integrals are evaluated as follows.

$$\int_{-\infty}^{+\infty} A(z,\omega-\omega_0)e^{-i(\omega-\omega_0)t}\frac{d(\omega-\omega_0)}{2\pi} = \tilde{A}(z,t) \tag{44}$$

$$\int_{-\infty}^{+\infty} (\omega-\omega_0)A(z,\omega-\omega_0)e^{-i(\omega-\omega_0)t}\frac{d(\omega-\omega_0)}{2\pi}$$

$$= \frac{1}{-i}\frac{\partial}{\partial t}\int_{-\infty}^{+\infty} A(z,\omega-\omega_0)e^{-i(\omega-\omega_0)t}\frac{d(\omega-\omega_0)}{2\pi} = i\frac{\partial}{\partial t}\tilde{A}(z,t) \tag{45}$$

$$\int_{-\infty}^{+\infty} (\omega-\omega_0)^2A(z,\omega-\omega_0)e^{-i(\omega-\omega_0)t}\frac{d(\omega-\omega_0)}{2\pi} = -\frac{\partial^2}{\partial t^2}\tilde{A}(z,t) \tag{46}$$

Equation (43) then becomes

$$\frac{\partial\tilde{A}}{\partial z} + k_1\frac{\partial\tilde{A}}{\partial t} + \frac{1}{2}ik_2\frac{\partial^2\tilde{A}}{\partial t^2} - i\Delta k_{NL}\tilde{A} = 0 \tag{47}$$

Introducing the retarded time τ as

$$\tau = t - \frac{z}{v_g} = t - ik_1 z \tag{48}$$

and we introduce the optical pulse defined by a function

$$\tilde{A}(z, \tau) = \tilde{A}(z, t) \tag{49}$$

Next the chain rule of differentiation is used to show that

$$\frac{\partial \tilde{A}}{\partial z} = \frac{\partial \tilde{A}_s}{\partial z} + \frac{\partial \tilde{A}_s}{\partial \tau}\frac{\partial \tau}{\partial z} = \frac{\partial \tilde{A}_s}{\partial z} - k_1 \frac{\partial \tilde{A}_s}{\partial \tau} \tag{50}$$

$$\frac{\partial \tilde{A}}{\partial t} = \frac{\partial \tilde{A}_s}{\partial z}\frac{\partial z}{\partial t} + \frac{\partial \tilde{A}_s}{\partial \tau}\frac{\partial \tau}{\partial t} = \frac{\partial \tilde{A}_s}{\partial \tau} \tag{51}$$

and analogously that $\frac{\partial^2 \tilde{A}_s}{\partial t^2} = \frac{\partial^2 \tilde{A}_s}{\partial \tau^2}$ These expressions are now introduced into Eq. (37), which becomes

$$\frac{\partial \tilde{A}_s}{\partial z} + \frac{1}{2} i k_2 \frac{\partial^2 \tilde{A}_s}{\partial t^2} - i \Delta k_{NL} \tilde{A}_s = 0 \tag{52}$$

The two equations obtained for the signal and pump beams in the case of two-beam coupling in a photorefractive material are, therefore,

$$\begin{aligned} \frac{\partial A_1}{\partial z} - i \Delta k_{NL} \tilde{A}_1 + \frac{i\beta_1}{2}\frac{\partial^2 \tilde{A}_1}{\partial z^2} &= 0 \\ \frac{\partial A_2}{\partial z} - i \Delta k_{NL} \tilde{A}_2 + \frac{i\beta_2}{2}\frac{\partial^2 \tilde{A}_2}{\partial z^2} &= 0 \end{aligned} \tag{53}$$

where

$$\beta_i = \left[\frac{d^2 k_i}{d\omega_i^2}\right], \omega_i = \omega_0 \tag{54}$$

with $i = 1, 2$ and

$$\Delta k_{NL} = \frac{n_2 \omega n_0}{2\pi}\left[|A_1|^2 + |A_2|^2 + \frac{A_1^* A_2}{1 + i\delta\tau}\right] \tag{55}$$

Substituting the value of, the Eq. (53) reduces to

$$\begin{aligned} \frac{\partial A_1}{\partial z} + \frac{i\beta_1}{2}\frac{\partial^2 A_1}{\partial z^2} - \frac{i n_2 \omega n_0}{2\pi}\left[|A_1|^2 + |A_2|^2\right] A_2 - \frac{i n_2 \omega n_0}{2\pi(1 + i\delta\tau)}|A_1|^2 A_2 = 0 \\ \frac{\partial A_2}{\partial z} + \frac{i\beta_2}{2}\frac{\partial^2 A_2}{\partial z^2} - \frac{i n_2 \omega n_0}{2\pi}\left[|A_1|^2 + |A_2|^2\right] A_1 - \frac{i n_2 \omega n_0}{2\pi(1 + i\delta\tau)}|A_2|^2 A_1 = 0 \end{aligned} \tag{56}$$

Equation (56) can be solved by the Fourier method. Approximating the expression $1 + i\delta\tau$ by one for $\delta\tau \approx 0$, and to achieve separation of variables, we consider the solution to these equations in the form

$$\tilde{A}(z, \tau) = \sqrt{I_{pi}} F_i(\tau) exp(i\Gamma_i z) \tag{57}$$

where Γ_i is defined as the wave propagation constant, $F_i(\tau)$ is a real function of τ. We set $\sqrt{I_{pi}} = 1$, where I_{pi} is the intensity of the pump beam.

Making a change of variables as in Eq. (17)

$$\lambda = \frac{2\Gamma}{\beta_i}, \mu = -\frac{2\gamma}{\beta_i} \tag{58}$$

where γ is the self modulation (SPM) parameter.

Again making further change of variables as

$$\sqrt{|\lambda|}\tau = T, F(\tau) = \frac{|\lambda|}{|\mu|} Q_i(T), Q_i^{'} = \frac{dQ_i(T)}{dT} \tag{59}$$

Equation (56) reduces to

$$\begin{aligned} Q_1^{''} + sgn(\lambda)Q_1 + sgn(\mu)A\left[Q_1^3 + 2Q_1Q_2^2\right] = 0 \\ Q_2^{''} + sgn(\lambda)Q_2 + sgn(\mu)A\left[Q_1^3 + 2Q_1Q_2^2\right] = 0 \end{aligned} \tag{60}$$

where

$$A = \frac{n_2\omega n_0\lambda}{\pi\beta_1\mu^2} \tag{61}$$

and sgn is the signum function with values ± 1. We can consider this as the Euler-Lagrange equation of a Hamiltonian system. Multiplying Eq. 60 by $Q_1^{'}$ & $Q_2^{'}$, respectively, and integrating with respect to time, the corresponding Hamiltonian is obtained:

$$H(Q_1, Q_1^{'}) = Q_1^{'2} + Q_1^2 + sgn(\lambda)\frac{A}{2}Q_1^4 + sgn(\mu)Q_1^2Q_2^2 = h \tag{62}$$

$$H(Q_2, Q_2^{'}) = Q_2^{'2} + Q_2^2 + sgn(\lambda)\frac{A}{2}Q_2^4 + sgn(\mu)Q_1^2Q_2^2 = h \tag{63}$$

where h is the constant of integration which can be obtained using boundary conditions. The behavior of the Hamiltonian dynamical system is investigated by considering the quantities λ, μ and h as parameters. When the parameters $\lambda < 0$and $\mu > 0$, Eq. (60) can be expressed as

$$Q_1^{'2} + Q_1^2 - \frac{A}{2}Q_1^4 + Q_1^2Q_2^2 = h \tag{64}$$

Applying the minimization criteria:

$$\frac{dh}{dQ'_i} = 0$$
$$\frac{dh}{dQ_i} = 0 \tag{65}$$

with $i = 1, 2$.

Using Eqs. (65), (64) can be expressed as

$$2Q_1\left[1 - A(Q_1^2 - 2Q_2^2)\right] = 0 \tag{66}$$

so that

$$Q_2^2 = \frac{Q_1^2 - 1}{2} \tag{67}$$

Substituting (66) in (67), the equation obtained is

$$Q_1'^2 + Q_1^2 - \frac{3A}{2}Q_1^4 + AQ_1^2 = 0 \tag{68}$$

Taking A=1, and multiplying by $\frac{2}{3}$ and integrating (68) the solution obtained is

$$Q_1 = -\frac{4}{3}sech\frac{2\sqrt{2}}{\sqrt{3}}T \tag{69}$$

Substituting for T from Eq. (59), the soliton solutions obtained is

$$Q_1 = -\frac{4}{3}sech\frac{2\sqrt{2}}{\sqrt{3}}\sqrt{\lambda}(t - k_1z) \tag{70}$$

When $(t - k_1z) \to \infty$, $A_1(z, t) \to 0$ and hence this is a bright solitary solution, a profile of which is shown in Fig. 1. Following the same procedure for Eq. (63), we get the solution as

$$Q_2 = -\frac{4}{3}sech\frac{2\sqrt{2}}{\sqrt{3}}\sqrt{\lambda}(t - k_2z) \tag{71}$$

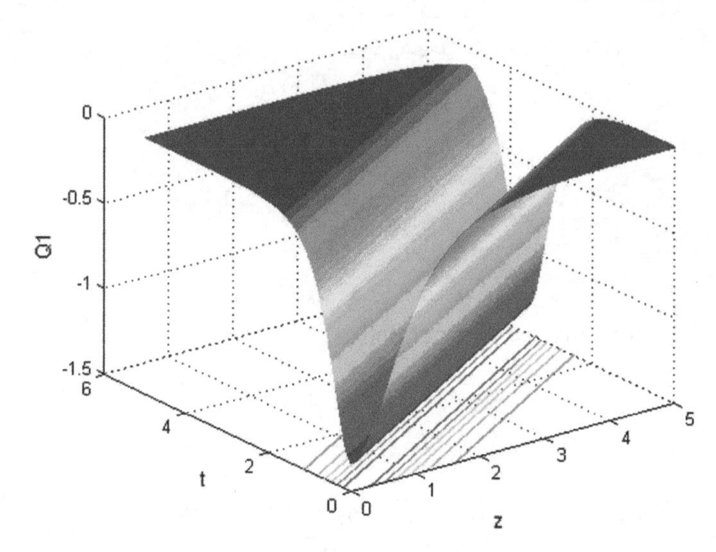

Fig. 1 Amplitude profile of bright soliton with $k_1 = 1.4$

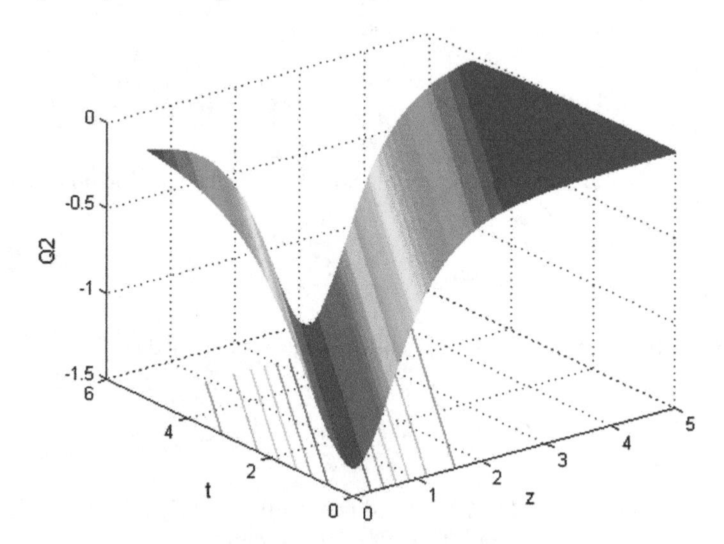

Fig. 2 Amplitude profile of bright soliton with $k_2 = 0.5$

8 Conclusion

In this article the nonlinear wave propagation through photorefractive media using Manakov model is investigated. The CNLSE is obtained which is solved and found the conditions under which soliton-waves can propagate. It is found that with an appropriate choice of experimental parameters, the input beam profile can be made to converge asymptotically to a soliton state whose PR nonlinearity compensate for

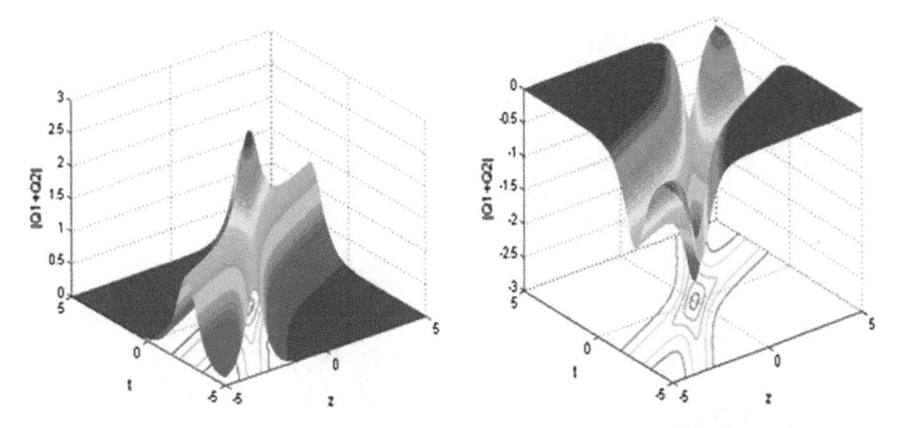

Fig. 3 Amplitude profile of bright soliton interaction and its inverted figure with $k_1 = 1.4$ and $k_2 = 0.7$

diffraction and the beam profile remains unchanged as the propagates. The profile of dark soliton solution which is obtained for $\lambda > 0$, $\mu > 0$ is given in Figs. 1 and 2 for different values of k_1. Figure 3 shows the interaction of two beams for certain values of k_1 andk_2. For all other cases ($\lambda < 0, \mu < 0$), ($\lambda > 0, \mu < 0$) and ($\lambda > 0, \mu > 0$), there are no solitary solutions. The results may find application in various fields of nonlinear optics, especially to study crystal orientation and beam polarization on photorefractive responses in the two-beam coupling configuration.

References

1. R.K. Bullough, P.J. Caudrey. *Solitons* (Springer, Berlin 1980)
2. M. Lakshmanan, S. Rajasekhar, *Nonlinear Dynamics* (Springer, 2003)
3. G.P. Agrawal, *Nonlinear Fiber Optics* (Academy Press, California, 2001)
4. A. Hasegawa, *Optical Solitons in fibers* (Springer, Berlin, 1980)
5. G.I. Stegeman, M. Segev, Special issue on Front. Optics (Rev.) **286**, 1518 (1999)
6. D.N. Christodulides, R.I. Joseph, Phys. Rev. Lett. A **141**, 37 (1989)
7. Hasegawa, Y. Kodama, *Solitons in Optical Communications* (Clarendon Press, 1995)
8. V.C. Kuriakose, K. Porsezian, Resonance, J. Sci. Educat. **15**, 643 (2010)
9. M. Segev, B. Crosignani, A. Yariv, B. Fisher, Phys. Rev. Lett. **68**, 923 (1992)
10. M. Segev, M. Shih, G.C. Valley, J. Opt. Soc. Am. B **13**, 706 (1996)
11. M.D. Iturbe-Castillo, P.A. Marquez-Aguilar, J.J. Sanchez-Mondragon, S. Stepanov, V. Vysloukh, Appl. Phys. Lett. **64**, 408 (1994)
12. M. Segev, B. Crosignani, A. Yariv, M.M. Feger, M. Bashaw, Phys. Rev. A **50**, 4457 (1994)
13. D.N. Christodoulides, M.I. Carvatho, Opt. Soc. Am. B **12**, 1628 (1995)
14. Z. Chen, M. Segev, T.H. Coskun, D.N. Christodoulides, Y.S. Kivshar, J. Opt. Soc. Am. B **14**, 3066 (1997)
15. M. Shih, M. Segev, G.C. Valley, G. Salamo, B. Crosignani, P. Di Porto, Electron. Lett. **31**, 826 (1995)

Neurons and Near-Death Spikes

Rose P. Ignatius

Abstract Near-death spikes or near-death surges represent sudden increase in neuron activity in the human brain before neurons end their firing. Just before a person is clinically dead, such spikes are observed in certain cases, so it got the name 'near-death spikes'. The reason for this behaviour is the lack of oxygen in brain. The neural network of the worm *Caenorhabditis elegans* resembles that of human brain. Hence it can be used to understand the simple dynamics of human brain. Within the network, the neurons are found to exhibit chaotic nature, even though their parameters are that of normal neurons. It is observed that when the strength of synaptic conductance is increased, initially the bursting synchronization, entropy of the network and the average firing rate decrease slightly and then increase. As the neurons of the network are made chaotic, 'near-death'-like surges of neuron activity are observed. Also, the brain dynamics changes from alert to rest state. It can be demonstrated that a particular type of noise called Lévy noise can generate 'near-death'-like surges in the neural network of the worm *Caenorhabditis elegans*. Identification of different parameter regions of Lévy noise at which the network makes transitions from one synchronous state to another and the mechanism behind them is a challenging subject. Such transitions are already reported in cortical regions of brain. The Lévy noise values at which the network displays generation of waves of different frequencies can be determined. This result suggests a new method for neuro stimulation in the case of traumatic brain injury. The neuronal network even displayed Gamma oscillations. If the parameters of the neurons are made chaotic, the network firing rate is diminished and it displayed Delta and Theta oscillations.

Keywords Neuronal network · Spatiotemporal pattern · Chaotic neurons · Lévy noise

R. P. Ignatius (✉)
Department of Physics, St. Theresa's College, Ernakulam, Kerala, India
e-mail: rosgeo@yahoo.com

Department of Physics, Al-Ameen College, Edathala, Aluva, Kerala, India

K. S. Sreelatha and V. Jacob (eds.), *Modern Perspectives in Theoretical Physics*,
https://doi.org/10.1007/978-981-15-9313-0_10

1 Introduction

Neurons communicate between each other through chemical and electrical synapses. Individual neurons show different firing patterns such as spiking, bursting, chaotic firing patterns and a mix of them [1]. When more than two neurons are coupled together, which is the case of real biological neuronal networks, their behaviour becomes still more interesting and complex due to the involvement of a large number of parameters pertaining to individual neurons, network and external factors. Such networks display complex phenomena like amplitude death, oscillation death, near-death-like spikes [2, 3], synchronization, coherence resonance, stochastic resonance, chimeras, spatiotemporal pattern formation and pattern selection [4].

In an undirected, homogenous *Caenorhabditis elegans (C. elegans)* neuronal network, due to the interplay of chemical and electrical synapses, chimera-like synchronization [5] pattern was observed. In a heterogeneous network of integrate-and-fire neurons, the macroscopic dynamics was found to be irregular, whereas the microscopic dynamics is linearly stable. Spatiotemporal activities of neuronal network are an important feature of brain dynamics like cognitive process, visual object detection, memory formation [6] and of seizures [7]. Parameters like coupling strength, network topology, network size, delays, type of neurons and external factors like electric fields and noise affect the spatiotemporal activities of neural networks [8, 9].

It was shown that single-neuron spiking in sensorimotor cortex of humans and monkeys can be predicted using the spiking history of small random ensembles. Interplay between time constants of active ionic currents and interspike interval causes differences in synchrony of excitatory networks, and hence, the difference in corresponding spatiotemporal pattern is formed.

Recently, it has been shown that a random perturbation at the boundary neurons induces spiral waves in a regular excitatory neuronal network. In another study, it was revealed that Lévy noise can induce the mode transition in firing activities of neurons under electromagnetic radiation.

Astrocytes are star-shaped glial cells of nervous tissues. Astroglia is the main power source of neurons. Astrocytes on receiving the neurotransmitters change the intracellular calcium concentration and hence modulate the electrical activities of neurons connected to it. Thus, they facilitate the communication between neurons. Firing patterns of neurons and spatiotemporal dynamics of neuronal network can be understood clearly from the dynamics of connected astrocytes. For example, the network of neuron–astrocyte displayed different modes of electrical activities under autaptic(self-feedback) driving, when calcium and inositol triphosphate concentrations of astrocytes are under control [10]. Seizure-like firings in neuronal network is attributed to dysfunction of neurons. It has been shown that astrocytes and neurons support each other in the propagation of seizure-like firings in neuronal network [11].

2 Spatiotemporal Activities of a Pulse-Coupled Biological Neural Network (PCNN)

We can study the spatiotemporal activities and effects of chaotic neurons on the dynamics of pulse-coupled biological neural network (BNN) [12]. BNN used in the study described here is that of the *C. elegans*. *C. elegans* is a soil worm. It is a simple multicellular organism for which the complete neural network is identified. Earlier studies reveal that human neuron networks are similar to that of *C. elegans* in terms of network motifs, types of synapses and neurons, etc. So the dynamics of *C. elegans* neuronal network can be translated to that of mammals [13]. Just like humans the nervous system of *C. elegans* contains both chemical synapses and gap junctions. Also, neurons in them are arranged as modules. Genome of *C. elegans* is 35% similar to that of humans. These properties make them prototypes for studying simple dynamics of human brain.

One-dimensional model of neural network can be used to study the dynamics of entire network of *C. elegans*. On application of sinusoidal external input to neurons, the dynamics for different values of external input and synaptic connections can be studied and it has been identified that the network exhibits fixed point, periodic, quasi-periodic and chaotic dynamics. The study presented here concentrates on spatiotemporal dynamics of the network and on the influence of chaotic neurons on it.

2.1 *Izhikevich Model of Neuron*

Izhikevich neuron is a simple model, capable of displaying wide range properties of real biological neurons. Computational minimalism of this model makes it suitable for large-scale network simulation. The neuron model is represented by the following equations [14].

$$w' = 0.04w^2 + 5w + 140 - k + I \tag{1}$$

$$k' = a(bw - k)$$

with the auxiliary after-spike resetting given as

$$\text{if } w \geq 30\text{mV},\ w \leftarrow c \text{ and } k \leftarrow k + d \tag{2}$$

where 'w' represents the membrane potential, k accounts for the activation of K+ ionic currents and inactivation of the Na^+ ionic currents. When membrane potential reaches the maximum value, 30 mV, the membrane potential and current are reset according to Eq. (2). The input current (synaptic current) to the neuron is delivered through the variable I. In Eq. (1), dimensionless parameters a and b represent the time scale of k and sensitivity of k to w, respectively. The dimensionless parameters c and d represent the after spike reset values of membrane potential w and membrane recovery variable k, respectively.

To study the network dynamics, the directed neural network of *C. elegans* is used. The adult *C. elegans* hermaphrodite has 302 neurons. Its neural network is organized into two, a large somatic nervous system (282 neurons) and a small pharyngeal nervous system (20 neurons). For the study presented here, the local network of 131 frontal neurons of the *C. elegans* is considered.

The network created is such that 99 neurons are made excitatory and the remaining 32 ones are made inhibitory interneurons. Excitatory neurons stimulate the neurotransmitters and increase the firing probability of other neurons connected to it (the feedback is chosen to be positive). But the inhibitory neurons inhibit the neurotransmitters. Parameters of the neurons are chosen such that they vary randomly from neuron to neuron.

The synaptic current to the ith neuron is described as follows:

$$I_i = I_{dc} + G_{syn} \sum_j A_{j,i} I_{fired} \tag{3}$$

where I_{dc} is the external dc pulse to the neuron, 'G_{syn}' is the total conductance of the synapse, j indicates the index of the fired neuron, A_{ji} is the (j, i)th entry of adjacency matrix A of the network and I_{fired} represents the unit dc feedback current from the jth fired neuron that is connected to ith neuron. I_{fired} assumes zero value in case of no firing. Thus the second term in Eq. (3) represents the total feedback from the network to the ith neuron. The inhibitory interneurons chosen are 7th to 13th, 27th to 40th, 76th to 79th, 82th to 86th, 124th and 125th neurons of adjacency matrix of *C. elegans.* Parameters of the neurons are chosen such that they vary randomly from neuron to neuron.

As the dynamics is examined, the average firing rate of the network fluctuates and that fluctuation increases with the increase in G_{syn}. It is verified that the fluctuation is present even when the parameters of all the neurons are same. But the amplitude of fluctuations increases as randomness of parameters is incorporated. With the increase in synaptic conductivity, the influence imparted on a particular neuron by other neurons of the network increases and hence average firing rate of the neurons of the network increases almost exponentially. This in turn increases the memory effect (long-term potentiation or depression) produced by the feedback current, and hence, fluctuations increase with the increase in strength of synaptic conductivity.

The parameters of individual excitatory neurons are selected such that they show any one of the behaviours such as regular spiking, intrinsic bursting or chattering. But when introduced in the network, all of them exhibit chaotic chattering behaviour regardless of their parameter values. The dynamics of inhibitory neurons do not show dependence on synaptic coupling, but the dynamics of excitatory neurons vary in accordance with synaptic conductance. Even for slight variation in input current [15] the neuron exhibits different spike patterns. Therefore, the neurons are very sensitive to slight variations in input current. To confirm the chaotic nature of neurons in the network, bifurcation diagram of a representative excitatory neuron can be drawn. In the network, the dynamics of individual neurons are controlled by network-related parameters rather than the individual parameter values of neurons.In pulse-coupled heterogeneous biological neural network, such an activity is not yet reported although the marked deviation of neuron activity in random and homogenous network from individual ones are already reported.

2.2 Entropy of Network

To measure the amount of information in the network, entropy can be computed. It determines the amount of information in the spike train of neurons. It is defined as

$$H = -\frac{1}{N}\sum_{i=1}^{N}\sum_{\Delta t} P_{ISI}^{i}(\Delta t)\log_2\left(P_{ISI}^{i}(\Delta t)\right) \tag{4}$$

where $P_{ISI}^{i}(\Delta t)$ is the probability of interspike interval of the ith neuron of the network, Δt is the bin size. To calculate the probability of interspike interval, for the small sample size, logarithmic binning method is used. The total number of bins is taken as 10 and the zeroth time is set as 1 ms.

The entropy of the studied network also first decreases slightly, remains almost constant for some values of synaptic conductivity and then increases exponentially with synaptic conductivity. The reason for this behaviour is that when the synaptic conductivity is low, the regular spiking behaviour of inter neurons dominates over the chattering behaviour of other neurons. As the connection strength between neurons increases, some of the excitatory neurons stops firing. For small values of synaptic conductivity, inter spike interval of the neurons is almost uniform. As the coupling strength between neurons further increases, more excitatory neurons fire and their chattering behaviour dominates over spiking of interneurons. Also the inter spike interval of the neurons becomes irregular or chaotic. This is the reason for enhanced entropy and asynchronous behaviour of the network. Thus, with the increase in strength of synaptic conductivity, entropy or chaos present in the network increases in a rapid manner. Such an irregular behaviour at high values of synaptic conductivity can be found in coupled neurons, in the mathematical model of homeostatic regulation of sleep–wake cycles, under high external input current and modulation noise [16].

2.3 Influence of Chaotic Neurons

To study the effect of chaotic neurons in the spatiotemporal dynamics of pulse-coupled neuronal network, neurons are made chaotic randomly step by step. Then the changes in the dynamics of network are studied by varying synaptic conductivity.

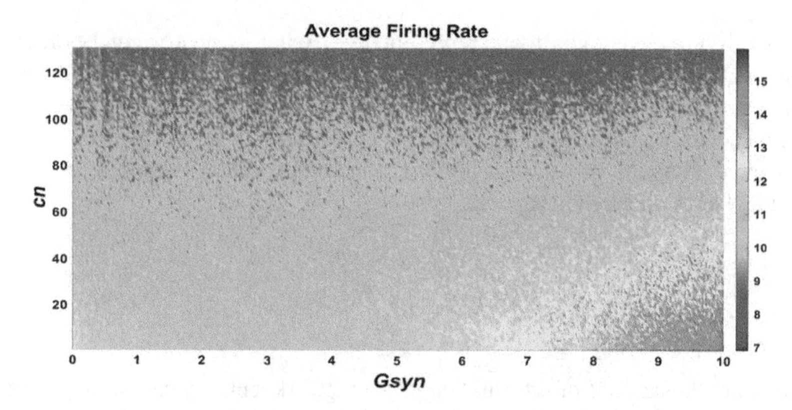

Fig. 1 Plot for average firing rate of the pulse-coupled network with chaotic neurons as a function of G_{syn}. When more neurons are made chaotic, alpha waves for which the average firing rate falls between 8 and 12 Hz dominates the network

Variation of average firing rate of the network with synaptic conductivity and the number of chaotic neurons is shown in Fig. 1. In the absence of chaotic neurons, for most of the studied values of synaptic strength, average firing rate is between 12 and 15 Hz, that is, the beta waves (12–30 Hz) dominate in the network. In humans, beta waves are associated with conscious state of mind, logical thinking, memory activation and alertness.

As neurons are randomly made chaotic, a shift toward alpha waves (8–12 Hz) is obviously visible in network activity. In humans, alpha waves correspond to resting state of brain, hence the effect of chaotic neurons is to change the brain state from alert to a meditative state. This is because of the change in behaviour of neurons from a spiking or chattering mode to a complete chattering mode and further to the inactive state as they advance in time (Fig. 2). Figure 1 indicates that network dynamics is independent of change in value of synaptic conductance when most of the neurons are chaotic. However, if the number of chaotic neurons in the network is small, when coupling or synaptic conductance strength is high, the beta waves seem to dominate (Fig. 1). Figure 2 displays the spatiotemporal activities of neurons for $G_{syn} = 10$ and $cn = 131$ at different values of the chaotic neuron number cn.

With the conversion of additional neurons into chaotic ones, the total spikes in the network decrease and finally the entire activity of the network stops. From Fig. 2, it is clear that all the neurons burst continuously for certain time just before terminating firings. In humans near the death, such a surge of neuron activity is detected before neurons end its firing [17]. This episodic firing in neuron is due to periodic propagation of calcium wave in astrocyte and persistent firing is due to homogenous calcium wave. As the parameters of the neurons are made chaotic, suppression of activities is observed (Fig. 2), or in other words, chaotic neurons are driven to firing death.

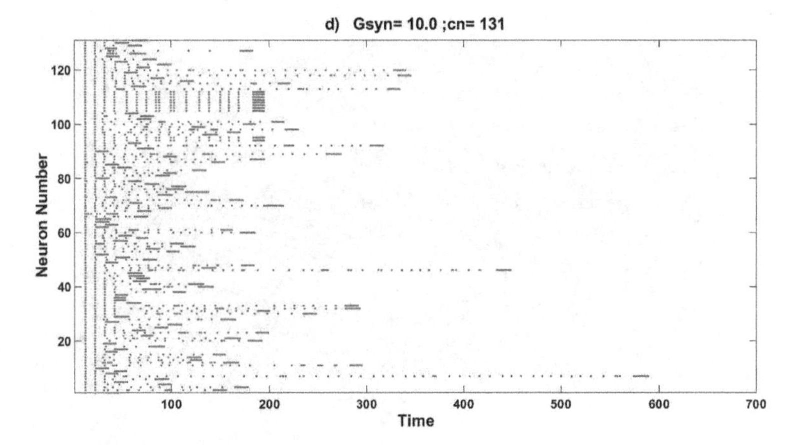

Fig. 2 Raster plots depicting firing times of PCNN at G_{syn} = 10, cn = 131. With increase in G_{syn} the synchronization in the network decreases. When more and more neurons are turned chaotic, their firing stops abruptly at around 56–600 ms

2.4 *Bursting Synchronization (B)*

B is a measure of degree of interspike distance synchrony. To find B, all the spike times of the network are arranged in the order of spike generation and then using the following equation bursting synchronization is calculated.

$$B = \frac{1}{\sqrt{N}}\left(\frac{\sqrt{\langle\tau_v^2\rangle - \langle\tau_v\rangle^2}}{\langle\tau_v\rangle} - 1\right) \tag{5}$$

where τ_v is the network interspike interval (ISI) between $(\tau_v + 1)$th and (τ_v)th spikes of the network, i.e., $\tau_v = \tau_{v+1} - \tau_v$. When B = 1, the network is in highly synchronous bursting state and low values of B indicates asynchronous behaviour.

As the number of the chaotic neurons and Gsyn increases, there is a tendency for the bursting synchronization of the network to become high. However, the increase is not uniform. For specific values of synaptic strength, when chaotic neurons in the network are above 120, the network moves toward complete synchronization (Fig. 3).

2.5 *Influence of Inhibitory Neurons*

Since all the interneurons of the network are not necessarily inhibitory, the above studies are repeated by making 32 randomly selected neurons in the network as inhibitory neurons. Then, the dynamics of both inhibitory and excitatory neuron can

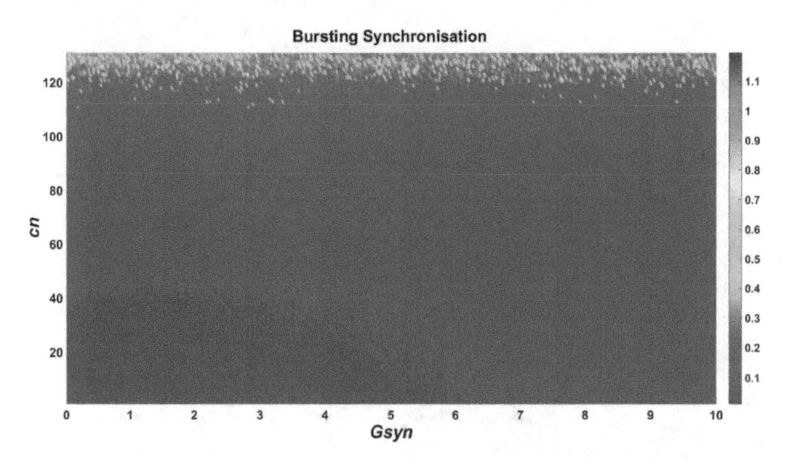

Fig. 3 Variation of bursting synchronization of PCNN with synaptic conductivity and number of chaotic neurons in the network. When most of the neurons are made chaotic, network displayed synchronization as the bursting synchronization value found to approach unity

be Fast Spiking (FS) or Chattering (CH) or Intrinsically Bursting (IB) or Regular Spiking (RS). In the case with the increase in synaptic strength more excitatory neurons of the network are fired more frequently. So average firing rate of the network increased slowly and then increased almost linearly with the increase in synaptic strength. Entropy of the system remained constant for very small values, then showed exponential growth and then increased linearly.

3 Lévy Noise-Induced Near-Death Spikes and Phase Transitions

Even after centuries of scientific search and validations, mankind still lacks the capacity to unlock the mystery behind death and related events. So near-death experiences always generate curiosity and are a subject of intense research. One such observation was reported that a surge in Electroencephalogram (EEG) activity near death and this phenomenon is called 'near-death surges' or 'wave of death'. One of their postulates is that, at near death, as the hypoxemia in patients increases beyond a threshold value, most of the neurons in the brain loses Na–K potential, which leads to a rise of electrical activity and a subsequent surge in EEG. Later studies verified their hypothesis.

In one such study, it is demonstrated that neurons take a delayed death compared to other organs and cells. During a near-death event, neurons end their firings with a burst. It is actually a sudden increase in neuron firings before the membrane voltage goes flat. They are linked with anoxia leading to sudden depolarization of membrane potential. This phenomenon is quite different from the synchronization induced spike termination (SIST) [18]. In SIST, a strong external impulse or a strong synaptic coupling in network of excitatory neurons caused hyper synchronization and subsequent cessation of neuronal firings. It is responsible for spontaneous termination of epileptic seizures. Whereas in a ‘wave of death’ hyper synchronization is not playing a role [19].

The analysis on ‘the spatiotemporal activities of a pulse-coupled biological neural network’ which discusses near-death spikes when the neurons are chaotic can be extended to explore the underlying reason for such behaviour, through dynamical analysis.

Under optimal conditions, noise can induce synchronization, coherence resonance, stochastic resonance and vibrational resonance in nonlinear systems [20]. It is demonstrated that Gaussian noise of particular intensity can induce spatial coherence resonance in a two-dimensional network of FitzHugh-Nagumo neurons. Fractional-order Gaussian noise can also enhance information carrying capacity of neurons. In another study, it is found that long-range connections can destroy noise-induced spatial coherence resonance (or cause decoherence). Influence of external electric or magnetic fields or both on the neuronal environment can also be treated as noise. Strong external electric field can influence the firing rate and synchronization of neural networks. Many theoretical and experimental studies have proved that both static and dynamic magnetic fields can suppress neuronal activities and cell growth [21–24].

Unlike Gaussian white noise, Lévy noise or Lévy distribution is characterized by small fluctuations, random large jumps and heavy tail. From predicting the path of a bacteria in a swarm to game theory [25], Lévy process finds applications in almost all fields of life, be it mathematics, finance, sociology, biology, physics, astronomy or meteorology. In finance, it is used to study the stock market, options pricing, etc. In physics, it finds applications in modelling turbulence, modelling chaos in Josephson Junction and in quantum group theory.

Like channel noise and membrane noise, there are different types of noises present in the neural system. Lévy noise is used to model the noisy environment of neuronal network [26]. Neuron models involving Lévy noise can display all the dynamical variations of membrane potential, be it diffusion, jumps in membrane voltage or jump-diffusion.

In a recent study, it was revealed that Lévy noise can improve electrical activity in a neuron under electromagnetic radiation. By adjusting appropriate parameters of noise, firing modes of the neuron can be varied. Influence of Lévy noise on the Izhikevich neuron, an all-to all coupled network and two excitatory coupled network are studied recently [27], and it is shown that when the Lévy noise's characteristic exponent is 0.5, the individual neuronal level firing activity and the network dynamics become irregular.

The study presented here gives the spatiotemporal activities of *C. elegans* neural network under the influence of Lévy noise [28]. It is relevant to find the influence of Lévy noise in ceasing or enhancing ‘wave of death’ activity and identify different parameter regions of noise at which the network makes transitions from one synchronous state to another and the mechanism behind them. Because during such transition period between asynchronous and synchronous firing states, network is more susceptible to changes in firing pattern of individual neurons. So, the result will help to control the dynamics of the neural networks in general.

The synaptic current to the ith neuron of the network is modified here as

$$I_i = I_{dc} + G_{syn} \sum_j A_{j,i} I_{fired} + \delta_i(t) \tag{6}$$

The Lévy noise received by the ith neuron is represented by 'δ_i'. 'δ_i' are random numbers following Lévy distribution given by Eq. (8). For that the required random numbers are sampled from continuous Lévy motion [29, 30]. The noise term, which are discrete numbers following Lévy distribution, is defined as

$$\delta_i(t) = \frac{dL(t)}{dt} \tag{7}$$

where $L(t)$ is the Lévy motion. Lévy noise can feature the small fluctuations and random large jumps of Lévy distribution. Where the α-stable Lévy process or Lévy distribution is represented as the characteristic function,

$$\phi_{(\theta)} = \begin{cases} \exp\left\{-D^{\alpha}|\theta|^{\alpha}\left[1 - i\beta\,\mathrm{sgn}(\theta)\tan\frac{\alpha\pi}{2}\right] + i\mu\theta\right\}, \alpha \neq 1 \\ \exp\left\{-D|\theta|\left[1 + i\beta\,\mathrm{sgn}(\theta)\frac{2}{\pi}\ln|\theta|\right] + i\mu\theta\right\}, \alpha = 1 \end{cases} \tag{8}$$

where 'α' is the stability parameter or characteristic exponent with values ranging from $0 < \alpha \leq 2$, 'β' is the skewness parameter having values ranging from $-1 \leq \beta \leq 1$, 'D' is the scale parameter and 'μ' is the mean of the distribution [31].

Influence of noise on the network dynamics is studied by varying α and D for $\beta = 0$ and $\mu = 0$. The values of β and μ are kept zero, because their values does not influence the network dynamics. Contribution of network to the observed dynamics is analyzed by varying the synaptic coupling constant G_{syn}. Further, the neuron parameters are selected such that the neuron behaviour changes to that of chaotic ones, as identified by Izhikevich and the influence of Lévy noise on the respective network is investigated. For the neurons 'I_{dc}' is selected so that the observed firing frequencies are comparable to that of humans.

To the best of our knowledge, no reliable mean field approximations are proposed for directed random networks. It is observed that even though the parameters of neurons, except of interneurons, are chosen randomly, they all showed identical spiking pattern. So, in effect there are two types of neuron firing patterns visible in the network vis. FS interneuron and Non-interneuron. Therefore, a representative neuron is chosen from each group and their bifurcation, phase space and change in spike pattern with noise parameters are explored to explain the observed network level dynamics.

For the firing rate F can have values ranging from 0.1 to 40 Hz and above according to the different mental processes in the case of humans. Delta waves (0.1–4 Hz)

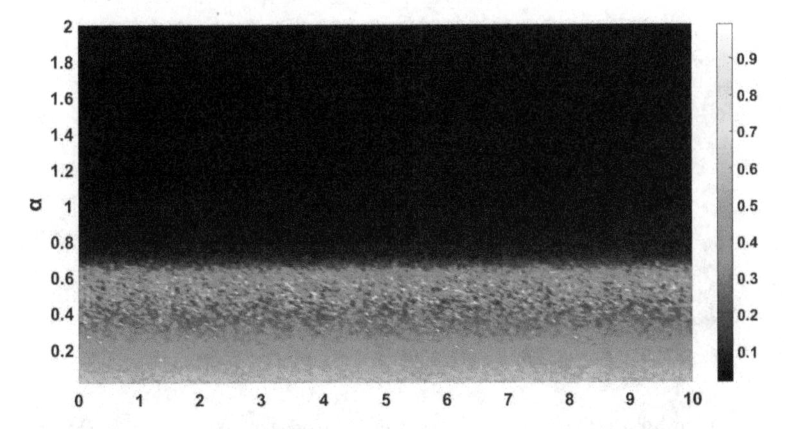

Fig. 4 Variations of busting synchronization of the network with α and G_{syn} at $D = 0.5$. The system displayed two states of synchronization with increase in α for a given synaptic coupling. The dynamics is independent of change in G_{syn} value. Similar results are got with α and D at $G_{syn} = 5.0$ (non-chaotic neurons)

corresponds to deep sleep or unconscious state. Theta waves range from 4 to 8 Hz indicating REM sleep or deep meditation. Alpha waves ranging from 8 to 12 Hz indicates relaxed or drowsy state of mind. Waves of frequency between 12 and 30 Hz called Beta waves correspond to alert state. Waves having frequency greater than 30 Hz are called Gamma wave, and they are associated with concentration, ecstasy or cognitive processes [32].

3.1 Bursting Synchronization in the Presence of Noise

It is found that the neuronal network remained unaffected by changes in G_{syn}, for all studied α values (Fig. 4). The neurons showed some amount of synchronization up to $\alpha \approx 0.7$. As the value of α increased further, the network changes to a complete asynchronous state. Some complete synchronized regions are also visible around $\alpha = 0.5$. For other values of D, the basic dynamics of the network remained essentially the same as that for $D = 0.5$.

3.2 Behaviour of the Network When Neurons are Chaotic

Significant changes in the network dynamics is observed only when the parameters of all the neurons are changed to that of chaotic neurons. When all the neurons are made chaotic at $\alpha = 0.5$, bursting synchronization among the neurons of the network

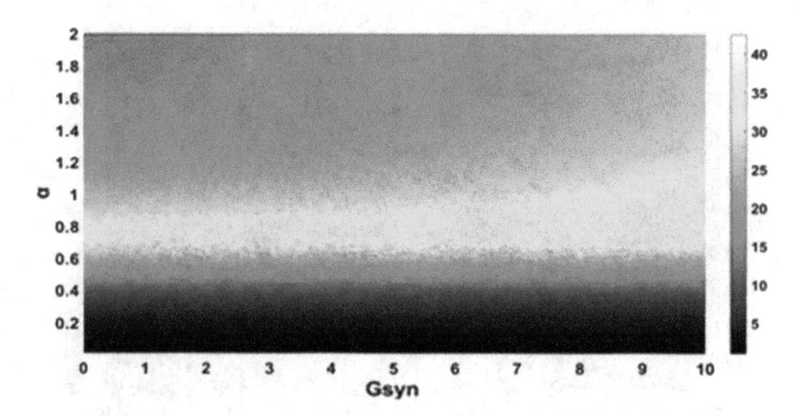

Fig. 5 Variations of average firing rate (Hz) of the network with α and G_{syn} at $D = 0.5$. With the increase in α, the network displayed a wide range of frequencies from delta to beta waves (non-chaotic neurons)

is decreased. As D and G_{syn} are varied, the complete synchronized regions found to be lost.

3.3 Average Firing Rate (F)

From Fig. 5, it is clear that the network displayed a wide range of frequencies as α is increased. For very low values of α (<0.1) network showed Delta waves (0.5–3 Hz). When α is gradually increased (but <0.8), the network activity is also changed to Theta (3–8 Hz), to Alpha (8–12 Hz) and finally to Hi-Beta range. When α is increased above 0.8, firing rate gradually decreased to low-Beta frequency. But high values of α causes an increase in average firing rate of the network, due to increased coupling strength. Since network displayed mainly two states of synchronization with increase in α value, a representative value for α is chosen from both of these regions and the response of the network towards variations in D and G_{syn} is investigated. From Fig. 6a, it is clear that Beta waves (waves of 12–40 Hz) dominate in the network. Average firing rate of the network is invariant with increase in G_{syn}. Whereas with increase in D, average firing rate gradually shifted from Hi-Beta to Lo-Beta, i.e. 24–12 Hz. For small values of D all the neurons in the network spikes irregularly (Fig. 7). They also displayed near-death spikes.

The behaviour for $\alpha = 0.8$ is very much different, because of the rich dynamics of noise at these values (Fig. 6b). The average firing rate exhibits an increase with respect to D and G_{syn}. The network displayed Gamma frequency (38–42 Hz) at these parameter regions.

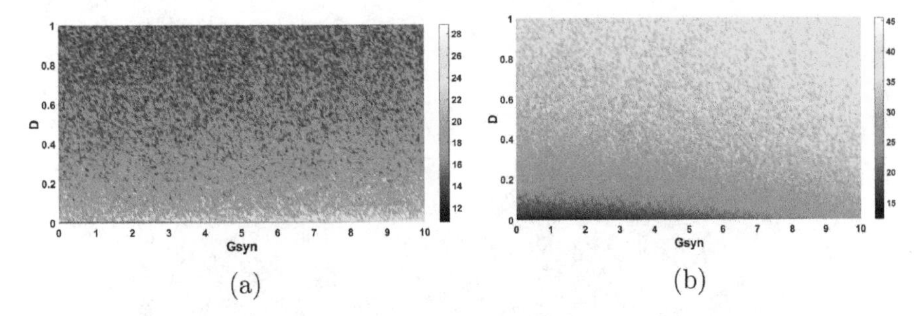

Fig. 6 Variations of average firing rate (Hz) of the network with D and G_{syn} **a** at $\alpha = 0.5$, beta waves dominate in the network. **b** At $\alpha = 0.8$, for small values of D the beta waves dominate in the network, but for large D values gamma waves appears (non-chaotic neurons)

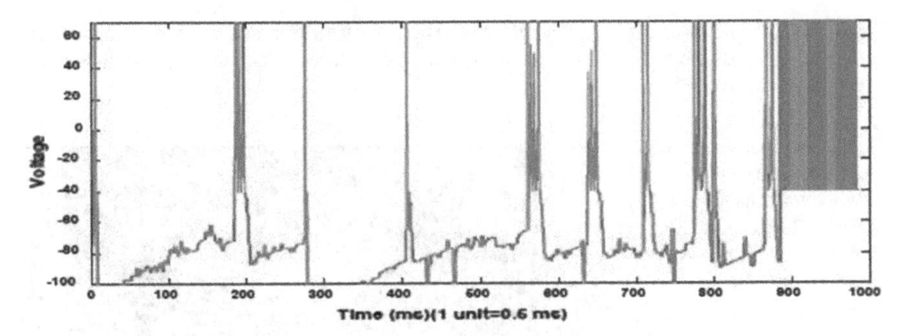

Fig. 7 Irregular spiking of a non-interneuron at $D = 0.05$, $G_{syn} = 5.0$, $\alpha = 0.5$, near-death spikes are also visible (neurons end its firing near t = 500 ms)

3.4 Average Firing Rate with the Chaotic Neurons in the Network

Here variations of average firing rate with increase in D, G_{syn} and α values are explored for the system, when all the neurons of the network hold chaotic parameters. For a particular G_{syn}, with increase in α value, 'F' is increased from Delta to Theta range, at $D = 0.5$ (Fig. 8). The average firing rate of the network shifted from Beta to Theta waves (firing decreased) at $\alpha = 0.5$ (Fig. 9). Here also the average firing rate is independent of G_{syn}, but it decreases with increase in D. When α is increased to 0.8, average firing rate is increased, but network dynamics remained the same as shown in Fig. 9. Then Theta and Alpha waves are visible in the network.

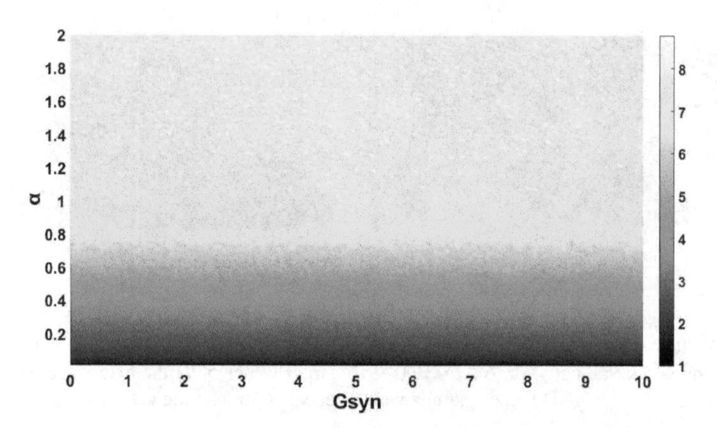

Fig. 8 Variations of average firing rate (Hz) of the network with α and G_{syn} when all the neurons of the network are chaotic at $D = 0.5$, network activity is independent of G_{syn}. But with increase in α average firing rate is increased

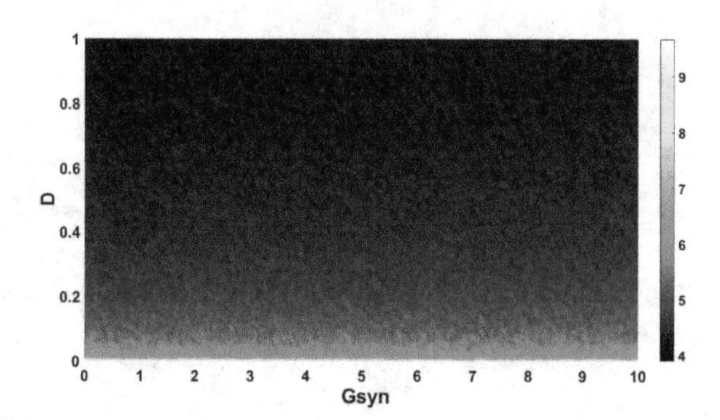

Fig. 9 Variations of average firing rate (Hz) of the network with chaotic neurons when D and G_{syn} are varied simultaneously, at $\alpha = 0.5$. Then the network displayed theta firings (3–8 Hz)

3.5 Variations of Neuron Firing with α

To study the variations of firing pattern of the neurons with α, raster plots are drawn. In Figs. 10 and 11, each dot indicates a spike. The network showed some amount of synchronization up to $\alpha \approx 0.7$. As the value of α is increased further, the network changes to a complete asynchronous state (Figs. 10 and 11).

From Fig. 10a, for low values of α, neurons fire synchronously but the oscillations within the network die out fast with a burst. The neurons of the network exhibited near-death-like spikes for α values less than or equal to 0.7 (Fig. 10a, b). With increase in α value, the firing of the network sustained longer (Figs. 10 and 11). Large values of characteristic exponent α of noise caused the asynchronous bursting within the network to sustain longer (Fig. 11) (Figs. 12 and 13).

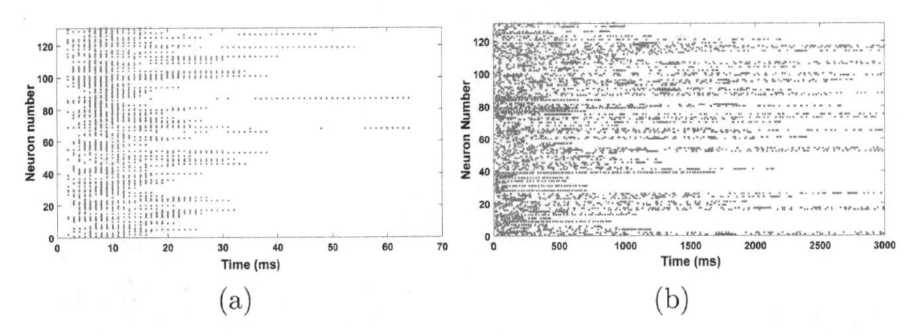

Fig. 10 **a** Raster plot of the network at G_{syn} = 5.0, D = 0.5 and $\alpha = 0.2$, each dot in the figure indicates a neuron firing. The network is displaying synchronization for a time period of about 64 ms and then the network dynamics is stopped and near-death spikes are visible. **b** At α = 0.7 the spikes are found to sustain for 3000 ms and only some of the neurons exhibit near-death spikes (non-chaotic neurons)

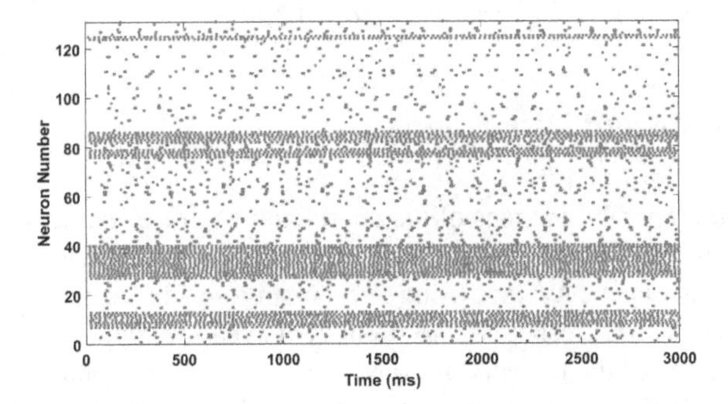

Fig. 11 Raster plot of the network at G_{syn} = 5.0, $D = 0.5$ and $\alpha = 1.5$. The network is displaying asynchronous bursting (non-chaotic neurons)

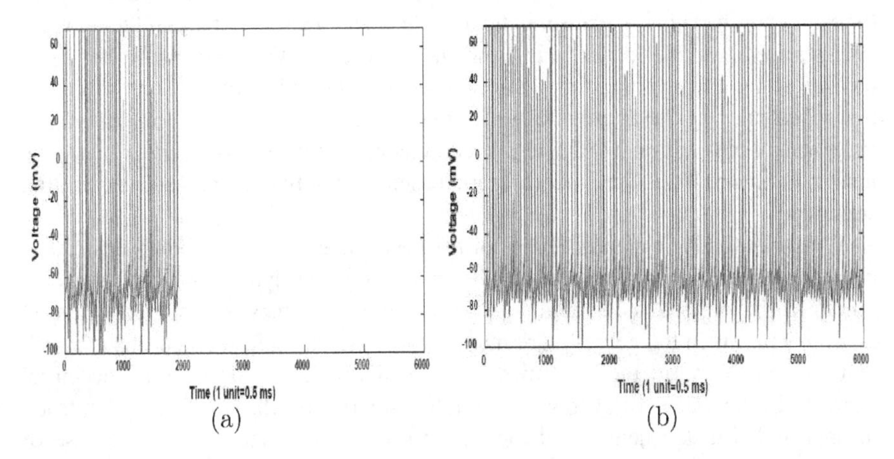

Fig. 12 **a** Irregular spiking of interneuron at $\alpha = 0.8$. The firings stopped around 2000 ms with a burst, indicating near-death spikes. **b** When α is increased to 1.8, inter neuron fired almost regularly

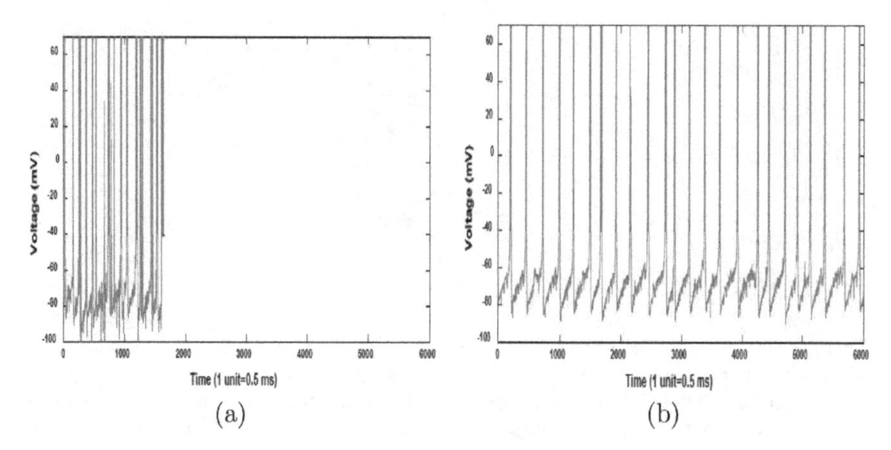

Fig. 13 For a non-inter neuron at $G_{syn} = 5.0$, $D = 0.5$, it indicates different modes of firing. **a** At $\alpha = 0.8$ neurons displayed irregular bursting, **b** at $\alpha = 1.8$, neuron bursting becomes almost regular

4 Conclusion and Discussion

For the spatiotemporal dynamics of the neural network of *C. elegans* it is observed that the non-chaotic network dynamics shows nonlinear dependence on the strength of synaptic conductance. Average firing rate, bursting synchronization of the network and entropy varies with respect to the strength of synaptic conductance.

As the neurons of the network are randomly made chaotic, interesting activities like increased alpha oscillations and 'near-death' like surges are observed. Average firing rate of the network is shifted from beta frequency range to alpha range as the neurons are turned chaotic. This indicated the change in brain state from alert state to resting phase. This is because, in the chaotic state, neuron activity in the network changes from spiking to chattering and finally to the inactive stage. All the three measures, namely, average firing rate, bursting synchronization and entropy show that when most of the neurons of the network are chaotic, the network activity is independent of the strength of synaptic conductance. But for small values of chaotic neuron number, all the three measures increased gradually with increase in coupling strength.

When randomly selected neurons of the network are made inhibitory(instead of all interneurons being inhibitory) and fast spiking, the dynamics of the network remained the same. A considerable change in the dynamics is observed when 32 random neurons in the network are made inhibitory. The selected neurons are such that they show any of the dynamics of FS or RS or IB or CH neuron, the initial decrease in the values of average firing rate, bursting synchronization and entropy disappeared. But the neurons of the network remained chaotic as in the case of network where all inter neurons are inhibitory and FS.

It can be concluded that within the network, excitatory neurons behaved differently compared to their independent activity. Normal neuron dynamics become chaotic within the network. When most of the neurons of the network are converted to chaotic ones, their activities become independent of each other.

The 'near-death'-like surges had not been reported earlier by the theoretical study of biological neuron network. The present study indicates that to soothe an overactive brain, some neurons of the network can be turned into chaotic. Further, experimental studies are to be performed before its use in treatment of diseases like stress, over anxiety, ADHD, Alzheimer's disease, etc. This study can be extended to the complete neural network of the *C. elegans*. A detailed mean field approximation of the present network can also be included in the future study.

This article also examines the influence of Lévy noise on the neural network of *C. elegans* through the dynamics of average firing rate and bursting synchronization. It is found that characteristic exponent of noise (α) influences the network dynamics more compared to scale parameter of noise (D) and coupling strength of synapse (G_{syn}). The present study identified that at α nearly equal to 0.7 the network makes transitions from one state of relatively high degree of synchronization to a state of almost complete asynchronization. It is known that with increase in α values, magnitude of noise changes from very high and random values to small and regular values. It is found that with increase in α values, the feedback current received by a neuron from the network peaks around $\alpha \approx 0.7$ and then decreases gradually. Consequently, for $\alpha < 0.8$, neurons displayed synchronizations and near-death spikes. With further increase in characteristic exponent of noise, firings of the neurons become asynchronous, sustains longer and the near-death spikes gradually get disappeared.

This work identifies the various parameter regions which helps in the control of the network dynamics. It is found that the network even displayed gamma oscillations for large values of α. The study also recognized α values, at which the network displayed the generation of waves of other frequencies. This result suggests a new method for neurostimulation in the case of traumatic brain injury [33].

In most of the studied parameter regions, noise enhanced bursting synchronization of the non-chaotic network of *C. elegans*. Compared to G_{syn}, D influences the network dynamics more. In the presence of Lévy noise, for some values of α, even the non-chaotic neurons of the network displayed near death like surges of firing. This observation is not yet reported in the literature.

Similar to that of the noise free case, chaotic neurons reduced the firing rate of the network from Beta range to Delta and Theta ranges. This reduction is a different feature in comparison to that of the non-chaotic ones for all values of α. The spiking rate and synchronization of the network are decreased due to decrease in value of average feedback current received by a neuron.

The results of the works support the observation of transitions between states of different degrees of synchronization in cortical regions of brain of rats. The model also enables to give the theoretical background for biological observation that during the transition period between asynchronous and synchronous firing state, network is more susceptible to changes in firing patterns of individual neurons.

References

1. E.M. Izhikevich, *Dynamical Systems in Neuroscience* (MIT Press, Cambridge, 2014)
2. W. Zou, D.V. Senthilkumar, A. Koseska, J. Kurths. Phys. Rev. E. **88**, 50901 (2013)
3. T. Banerjee, D. Biswas, Chaos Interdiscip. J. Nonlinear Sci. **23**, 43101 (2013)
4. J. Ma, J. Tang, Sci. China Technol. Sci. **58**(12), 2038 (2015)
5. C.G. Antonopoulos, A.S. Fokas, T.C. Bountis, Eur. Phys. J. Spec. Top. **225**(6–7), 1255 (2016)
6. W. Sato, T. Kochiyama, S. Uono, Sci. Rep. **5**, 12432 (2015)
7. R.M. Cichy, A. Khosla, D. Pantazis, A. Torralba, A. Oliva, Sci. Rep. **6**, 27755 (2016)
8. H. Wang, Y. Chen, Nonlinear Dyn. **85**(2), 881 (2016)
9. J. Wu, Y. Xu, J. Ma, PLoS One **12**(3), e0174330 (2017)
10. S. Guo, J. Tang, J. Ma, C. Wang, Complexity **2017**, 4631602 (2017)
11. J. Tang, J. Zhang, J. Ma, G. Zhang, X. Yang, Sci. China Technol. Sci. **60**(7), 1011 (2017)
12. K.K. Mineeja, R.P. Ignatius, Nonlinear Dyn. **92**(4), 1881 (2018)
13. M. Schröter, O. Paulsen, E. Bullmore, Nat. Rev. Neurosci. **18**, 131 (2017)
14. E.M. Izhikevich, IEEE Trans. Neural Netw. **15**, 5 (2004)
15. S. Nobukawa, H. Nishimura, T. Yamanishi, J. Liu, *Emerging Trends in Computational Biology, Bioinformatics and Systems Biology* (Elsevier Inc., 2015), p. 355
16. W. Jin, Q. Lin, A. Wang, C. Wang, Complexity **2017**, 4797545 (2017)
17. L. Norton, R.M. Gibson, T. Gofton, C. Benson, S. Dhanani, S.D. ShemieL, R. Ward Hornby, G.B. Young, Can. J. Neurol. Sci. **44**(2), 139 (2017)
18. M. Uzuntarla, J.J. Torres, A. Calim, E. Barreto, Neural Netw. **110**, 131 (2019)
19. B.J. Zandt, B. ten Haken, J.G. Van Dijk, M.J.A.M. van Putten, PLoS One **6**(7), e22127 (2011)
20. F.Y. Gao, Y.M. Kang, X. Chen, G. Chen, Phys. Rev. E. **97**(5), 1–11 (2018)
21. J. Aguila, J. Cudeiro, C. Rivadulla, Cereb. Cortex. **26**(2), 628 (2016)
22. A.D. Rosen, PIERS Online **6**(2), 133 (2010)
23. J.J. Gonzalez-rosa, V. Soto-leon, P. Real, C. Carrasco-lopez, G. Foffani, B.A. Strange, A. Oliviero, J. Neurosci. **35**(24), 9182 (2015)
24. A. Oliviero, L. Mordillo-Mateos, P. Arias, I. Panyavin, G. Foffani, J. Aguilar, J. Physiol. **589**(20), 4949 (2011)
25. G. Ariel, A. Rabani, S. Benisty, J.D. Partridge, R.M. Harshey, A. Be'er, Nat. Commun. **6**, 8396 (2015)
26. R. Cai, X. Chen, J. Duan, J. Kurths, X. Li, J. Stat. Mech. **063503**, (2017)
27. M. Vinaya, R.P. Ignatius, Nonlinear Dyn. **94**(2), 1133 (2018)
28. K.K. Mineeja, R.P. Ignatius, Nonlinear Dyn. **99**, 3265 (2020)
29. J. Yang, W. Zhou, P. Shi, X. Yang, X. Zhou, H. Su, Neurocomputing **156**, 231 (2014)
30. L. Zhou, Z. Wang, J. Zhou, W. Zhou, Neurocomputing **173**(3), 1235 (2016)
31. Y. Xu, Y. Li, H. Zhang, X. Li, J. Kurths, Sci. Rep. **6**, 31505 (2016)
32. Y. Buskila, A. Bellot-Saez, J.W. Morley, J.W. Front, Neuroscience **13**, 1125 (2019)
33. A. Pevzner, A. Izadi, D.J. Lee, K. Shahlaie, G.G. Gurkoff, Front. Syst. Neurosci. **10**, 30 (2016)
34. C. Zeng, Q. Yang, Chaos **25**, 123114 (2015)

Machine Learning and the Bigdata Paradigm

Ninan Sajeeth Philip

Abstract The use of machines to replicate human intelligence so that they may be used to consistently predict scenarios without human intervention is the subject of machine learning. The article describes two of the machine learning models that were developed by the author during his Ph.D. under the supervision of Prof. K Babu Joseph. The methods have high relevance in the present era that witnesses data explosion in all branches of knowledge, nicknamed Bigdata.

Keywords Machine learning · Neural networks · Bigdata

1 Introduction

The human brain works with the help of chemical computers called Neurons. Each neuron has a set of output called axons and inputs called dendrites. [See Fig. 1]. The output from the axons are input to the next set of dendrites and the strength of the connection between them is controlled by the concentration of the chemicals exchanged between them. This model of neurons lead to the design of computer-based neurons that also can accept input and pass on outputs to subsequent compute nodes. A network of such neurons is called an Artificial Neural Network or ANN. The experiential learning that takes place in the human brain is still not fully understood. But the process of learning through examples in the initial childhood years is well understood [1]. During this time, several chemical changes happen in the brain and the neurons learn to fire in accordance with every stimulus such that the same amount of chemicals are fired when a similar event occurs. The ANN mimics the chemical transportation strength in the biological neuron by assigning a connection weight between the output node and the subsequent input nodes. A set of examples are then used to adjust these connection weights so that the output of the ANN matches the

N. S. Philip (✉)
Artificial Intelligence Research and Intelligent Systems, Thelliyoor, Kerala, India
e-mail: ninansajeethphilip@airis4d.com

Department of Physics, St. Thomas College, Kozhencherry, Kerala, India

K. S. Sreelatha and V. Jacob (eds.), *Modern Perspectives in Theoretical Physics*,
https://doi.org/10.1007/978-981-15-9313-0_11

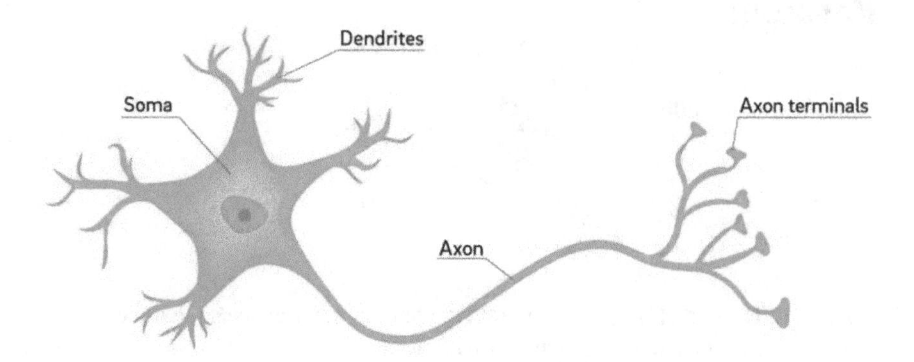

Fig. 1 The Biological Neuron takes input from external stimulus through its dendrites and the processed output is taken from the neuron to other neurons through the axon and chemical transporters between the axon endings and dendrites of other neurons

scenario described by the input. This process of updating the connection weights is what is called training. Like the biological neuron, a trained ANN will respond with the same output when similar inputs are applied to it.

Although the possibility was well understood and was experimented in different ways, it turned out to be impossible to adjust the weights except in trivial problems until the backpropagation algorithm was developed [2]. Backpropagation algorithm minimises the overall prediction error of the ANN by moving the weights along the negative gradient of the error surface. To understand what this means, assume that there is some value of the weight that correctly predicts the output. If the weight is increased or decreased, there will be an error which will be negative on one direction and positive on the other. Squaring the error makes it positive on either side from the optimal value giving the error surface a parabolic shape. If random initial weights are assigned and predictions are made, some will be predicted correctly while some others will fail. In the case of all the failed predictions, their weights will be somewhere on the parabola on either side of the optimal value for that weight. [See Fig. 2] The gradient at the optimal location marked as the global cost minimum location will be zero (because derivative at the point with respect to change in w will be zero). Also the gradient is positive on the RHS (rising) of the minimum and it is negative at the other side (falling). Thus, moving along the direction of the negative gradient will always move in the direction of the minimum. That means, gradually increasing or decreasing the weights (W) proportional to the negative of the gradient will help to reach the minimum. Since we are climbing down the gradient in both cases, it is called Gradient Descent Algorithm. The possible architecture of an ANN is shown in Fig. 3. Each of the pink balls represent an input node of the ANN. Traditionally, each such input is called a feature. For example, in a problem where the goal is to differentiate apples from oranges, colour, shape, texture, etc., can be

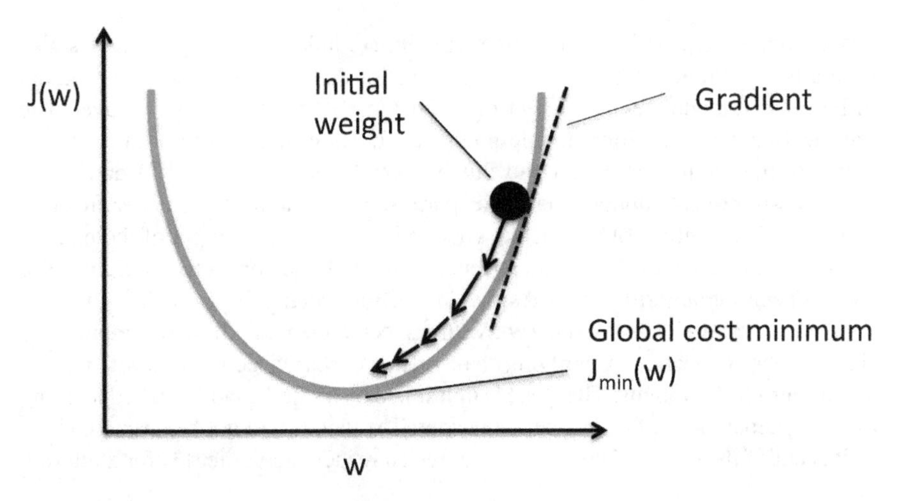

Fig. 2 The correct output is produced by the network when the weight W is having the value at the point shown by J_{min} [10]. The error J(w) is high when the weight is either higher or lower than W at J_{min}

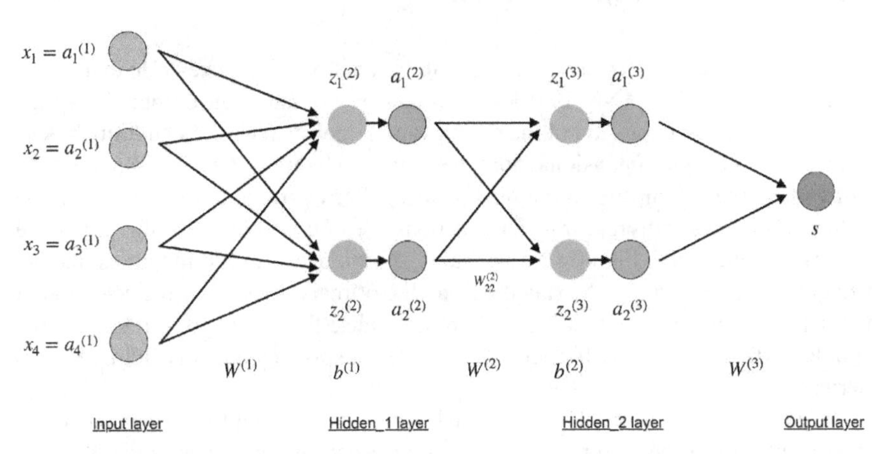

Fig. 3 A typical ANN is shown. The input nodes are shown in pink and the output node is shown in blue. There could be multiple output nodes too. The nodes between the input and output nodes are called hidden nodes and they have an associated nonlinear transfer function, also known as activation function, shown as green in the diagram [12]

the input features. A representation in which each feature followed by the label of its kind is called a feature vector. Intuitively, a feature vector points in the direction of the object in a feature space with the features as axis.

Mimicking the biological neuron, there could be several neurons and their connection before reaching the output node is shown as blue in Fig. 3. The nodes between input layer of nodes and output (layer) node(s) are known as hidden nodes and their layers are called hidden layers. They are called so because there is no access to these nodes from outside and everything about them are hidden. The hidden nodes also have a transfer function (also known as activation function and is shown as green in Fig. 3) that modulates the output before passing it to the subsequent layer. The transfer function is a differentiable nonlinear function that also re-normalises the output from each hidden node to a value between 0 and 1 or -1 and $+1$ depending on the design of the model. The major purpose of the nonlinear transfer function is to introduce nonlinearity in the network and to prevent the output of the node to explode to large values due to the multiplication with the connection weights. The backpropagation algorithm updates the connection weights by a small amount by considering the derivative of the error with respect to the input. It then computes the slope at each connection by applying chain rule on the partial derivatives with respect to the connection weights. The slope is multiplied by a small quantity called learning rate which increments the connection weight. The details will not be discussed here as it is out of the scope of this article. Interested readers may refer [3] for a tutorial.

1.1 Adaptive Transfer Function

The transfer function, also called the activation function, plays a key role in the nonlinear behaviour of the ANN. Besides allowing to re-normalise the output of the node to prevent factorial explosion of the output values, they also work as building blocks to construct the high-dimensional nonlinear decision hyper surface that separates the different classes of entities in the feature space. For the same reason, based on the nature of the data, different transfer functions are in use to improve the predictive accuracy of the ANN. Figure 4 gives a table of various activation functions that are used in machine learning. As stated before, the primary criteria for the selection of the activation function is that it should re-normalise the output from the node and should be differentiable. Differentiable because backpropagation algorithm requires derivatives to update the weights.

In 2002, Ninan and Joseph [4] proposed an adaptive transfer function that updates its structure during the training process to optimise both learning speed and prediction accuracy of the ANN. For this, the $tanh(x)$ function was generalised to read as

$$ABF = \frac{a + tanh(x)}{1 + a}$$

with a as a free adjustable parameter. The modification was shown to improve the learning speed and accuracy of the trained network. In a later study [5], they demonstrated that the rainfall pattern in Trivandrum over the 87 years prior to 2003 could be reliably predicted using ANN using adaptive transfer functions. This work got lot of attention because it demonstrated the predictability of monsoon rains in Trivandrum district prior to the dates when the chaotic changes due to global warming was not visible.

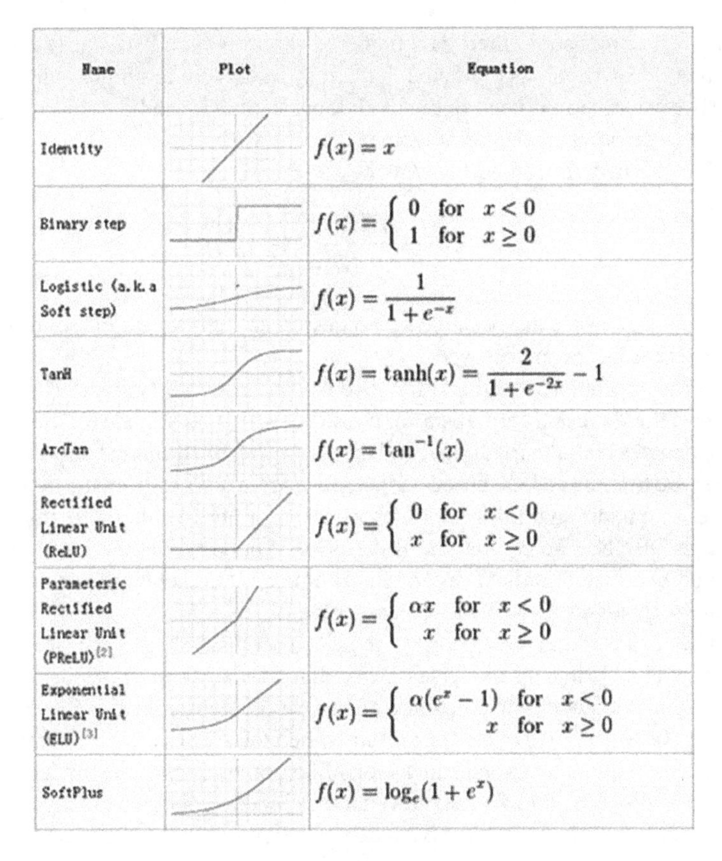

Name	Plot	Equation
Identity		$f(x) = x$
Binary step		$f(x) = \begin{cases} 0 & \text{for } x < 0 \\ 1 & \text{for } x \geq 0 \end{cases}$
Logistic (a.k.a Soft step)		$f(x) = \frac{1}{1+e^{-x}}$
TanH		$f(x) = \tanh(x) = \frac{2}{1+e^{-2x}} - 1$
ArcTan		$f(x) = \tan^{-1}(x)$
Rectified Linear Unit (ReLU)		$f(x) = \begin{cases} 0 & \text{for } x < 0 \\ x & \text{for } x \geq 0 \end{cases}$
Parameteric Rectified Linear Unit (PReLU) [2]		$f(x) = \begin{cases} \alpha x & \text{for } x < 0 \\ x & \text{for } x \geq 0 \end{cases}$
Exponential Linear Unit (ELU) [3]		$f(x) = \begin{cases} \alpha(e^x - 1) & \text{for } x < 0 \\ x & \text{for } x \geq 0 \end{cases}$
SoftPlus		$f(x) = \log_e(1 + e^x)$

Fig. 4 Transfer function, also called the activation function, can be any function that is differentiable and has a finite output range [12]. The table shows a set of possible activation functions and their plots

1.2 The Difference Boosting Neural Network

While ANN is a great idea and a breakthrough technology, it is not guaranteed to give the best results. This is mostly because there could be multiple minima in the error surface and instead of reaching the global minima, the weights might get trapped elsewhere that may give sub-optimal performance. Moreover, the iterative training process may take a long time to converge, thus adding to the computational overhead.

A more profound model is the Bayesian formalism that relies on likelihood and circumstantial knowledge for making efficient decisions. The likelihood is the experiential knowledge about the fraction of times a feature observation has resulted in a particular event. For example, clouds may cause it to rain and the likelihood is the fraction of times observation of clouds resulted in rain. Circumstantial knowledge is also a consideration in Bayesian formalism. It is called the prior. It is the background information about the probability of the event to happen in a given situation. For example, if it is monsoon, there is a higher probability clouds to cause rain. Or if the weather department has predicted that it may rain, there is a higher chance for it to rain. The prior is not related to the likelihood, though it can be used to weigh the likelihood to favour a particular outcome.

The Bayes Theorem can be stated as

$$P(A|B) = \frac{P(B|A) \times P(A)}{\sum_j P(B|A_j) \times P(A_j)}$$

where A_j represent all the possible outcomes including the test outcome A given that the event B has been observed.

Ninan and Joseph proposed a Difference Boosting Neural Network (DBNN) [6] that follows the Bayesian formalism for constructing the neural network model. The DBNN estimates both the likelihood and the prior from the training data. This means that the entire training is done based on the information derived from the training data and there is no additional information on either prior nor likelihood that is required to train the DBNN. The estimation of the prior is done with an adjustable weight that is modified using gradient decent algorithm and the likelihood is computed by counting the incidents that favour each outcome.

The DBNN has found wide application in Astronomy and Astrophysics. With the discovery of Hubble that the universe is expanding, it was realised that our Milky Way is just one of the innumerable collection of galaxies in the visible universe formed by clusters of 10^9 to 10^{11} or more stars. Ground based telescopes are not able to resolve the stars in nearby galaxies and they looks like clouds in the viewfinder. For this reason, for a long time they were believed to be clouds in our own galaxy. Although modern telescopes are able to resolve much better than the 2 m class telescope used by Hubble, it is still not easy to resolve and visually state whether an object is a star or a galaxy. This is because the farther a galaxy is, lesser will be its visual brightness and so they might appear like stars (point sources). However, with the help of dedicated software tools such as IRAF, it is possible to visualise the non-Gaussian profile of the light beam that indicates that these objects are not point sources. Needless to say, this is a tedious procedure and is not practical when surveys are conducted.

Hubble Deep Field Survey took longtime exposure of a few regions of the sky outside the field of view of the Milky Way galaxy. Ninan and Joseph used DBNN to accurately classify stars and galaxies [7] in Hubble fields with accuracy comparable to that made by human experts. While the human expert might take weeks to do the classification, DBNN could do it in a fraction of a second.

After stars and galaxies were identified, the next attempt was to estimate the population of different type of stars and galaxies. Though morphological structures of celestial objects can be confirmed through imaging, the confirmation of their nature and properties can be done only through spectroscopy. However, since the objects are very faint compared to the brightness of the background sky, spectroscopy is possible only on a few bright objects. Even for them, because spectra is taken by scattering the incident light across a wider area that represent the spectrum, long time exposures are required to obtain reliable spectra. These are major limitations of spectroscopic surveys.

When spectroscopic surveys were done, it was noted that there were a few objects that looked like stars but had a spectra that was completely different from any known spectra of stars. Since they looked like stars in the imaging surveys, they were named quasi-stellar objects or Quasars in short. Later studies revealed that quasars are not stars but are the active nucleus of distant galaxies. Because the nucleus is so bright compared to the rest of the galaxy, they could outshine the galaxy and appear as point sources. The Sloan Digital Sky Survey (SDSS) took the theme of doing spectroscopic confirmation of quasars for furthering the scientific understanding of the process. They could spectroscopically confirm about 120 thousand quasars, 930 thousand galaxies and 460 thousand stars in their data release DR7. Sheelu Abraham and Ninan [8] used DBNN to apply the spectral information on the five band imaging servery conducted by SDSS to identify candidates for spectroscopic confirmation. They used nine different image magnitudes and i band image magnitude to construct a ten-dimensional feature space for all the spectroscopically confirmed objects. Since the goal of the study was to accurately identify quasar candidates, a small region of the feature space where most quasars are found was selected for the study. A subset of about 14 thousand objects from the region was used for training the DBNN and the remaining were used for testing the accuracy of the predictions. The test results are given in Fig. 5 After confirming the reliability of the model, they went on to produce a photometric catalog of 119 thousand possible Quasars from the same region. While SDSS took 15 years of observations to produce the spectroscopic catalog of about 87 thousand, the identification of 119 thousand new candidates took only less than a minute for labelling. More than 99% of the predicted quasars were spectroscopically confirmed in subsequent spectroscopic surveys by SDSS.

Qusars house super massive blackholes that are vigorously consuming gas, dust and even nearby stars in the host galaxy. Part of the energy that is released from the accretion disc that spins round the blackhole is what outshines the galaxy and appear as quasars. An even higher release of energy occurs when neutron stars or blackholes merge to form a single blackhole. It could be so huge that it may produce gravitational waves that may propagate as space-time fluctuations across the visible universe. The gravitational wave produces orthogonal modulations in space time

Object	DBNN Predictions					
Type	Star	Galaxy	Quasar	Star-Late	Completeness	Contamination
Star	18,337	0	90	0	99.51 %	0.47 %
Galaxy	27	705	120	0	82.74 %	0.00 %
Quasar	34	0	86,852	0	99.96 %	0.28 %
Star-Late	25	0	34	242	80.40 %	0.00 %
Total	18,423	705	87,096	242	99.69 %	0.31 %

Fig. 5 The DBNN could correctly identify over 99% of the Quasars without increasing contamination around 0.31%

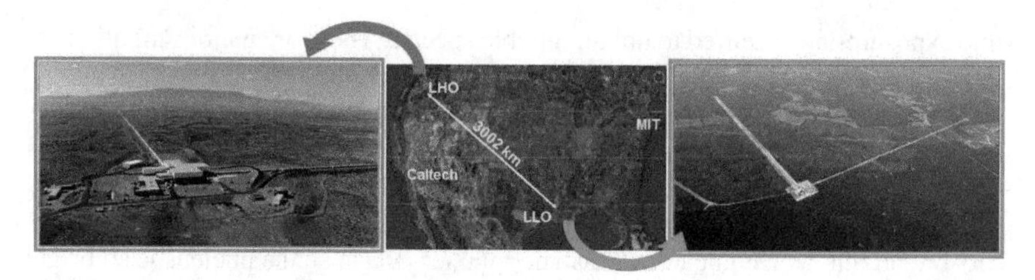

Fig. 6 The LIGO gravitational wave detector is a technological marvel that works like a Michelson Interferometer that has arm lengths of about 4 km each. There are two detectors of similar type, LIGO Hanford in southeastern Washington State and LIGO Livingston that is 3002 km away [13]

that will appear as changes in the length of the arms of an interferometer kept in its path. The changes in arm length result in changes in the fringe patterns that can be deconvolved to reconstruct the nature of the source of the gravitational wave (Fig. 6).

The Laser Interferometer Gravity Wave Observatory (LIGO) operated by Caltech and MIT consists of two similar interferometers separated by a distance of 3002 km. Each arm of the interferometer is about 4Km long and is sensitive enough to detect the vibrations produced by trucks moving miles away from its location. Such arm lengths are required to produce constructive or destructive interference by the oscillations produced by a passing gravitational wave. Two interferometers are used since a genuine gravitational wave should produce same patterns in both, delayed by the speed of light to reach from one to the other. This way it is possible to isolate fringe changes due to local disturbances such as moving trucks, foot steps of people or animals, earthquakes, etc. To improve sensitivity and to reduce the effect of the surrounding disturbances, the arms of the interferometers are evacuated and the mirrors are made to virtually float in the tube. In spite of all these precautions, several kinds of noise get into the detector and are in general called glitches. Glitches and actual signals are transients, meaning that they appear and disappear after a short time, see Fig. 7. Though it might appear from Fig. 7 that detection and isolation of glitches might be a straightforward problem, removal of glitches from real signal is extremely difficult due to the poor signal-to-noise ratio (SNR) of the detector. It happens that the glitches are concealed in the inherent noise due to various other sources making it invisible even for the expert eye, see Fig. 8. The LIGO detector usually have SNR ranging from 10 to 100 and detection at low SNR is really challenging. However, since signal detection and isolation is the sole purpose of the detector, transient classification is an indispensable part of the gravitational wave detector pipeline. Two postdoctoral students of IUCAA, Nikhil and Sheelu [9] observed that though the transients hidden in noise are apparently invisible, they become visible if the wavelet transform of the signal is taken. Since wavelet transform has good time resolution, it is possible to identify the exact location of the transient, see Fig. 9. They also noted that the wavelet energy for each transient is distinct that the nature of the transient can be accurately determined from the detailed coefficient wavelet energy patterns, see Fig. 10. The next goal was to develop a machine learning tool

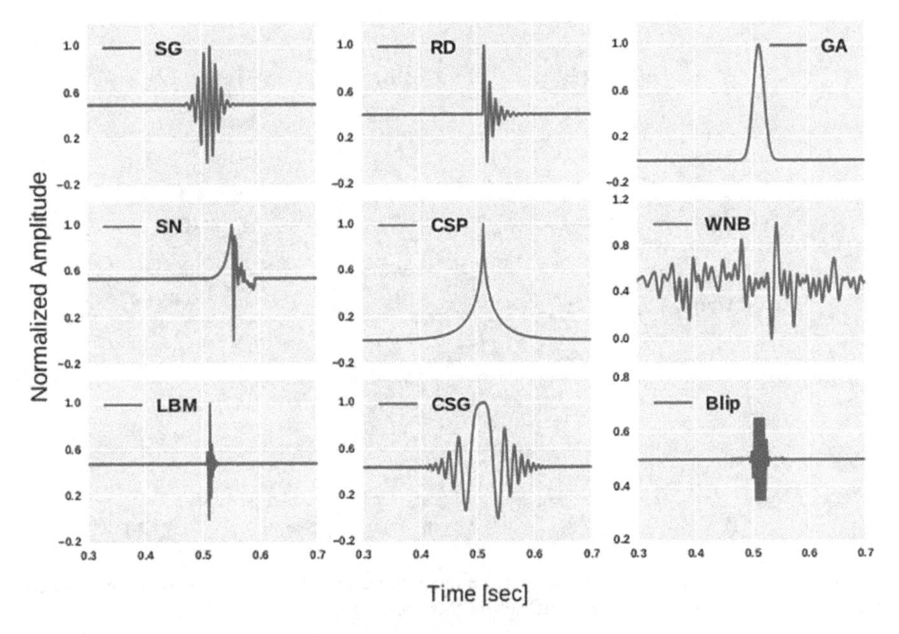

Fig. 7 Different type of glitches observed in gravitational wave detectors. Each of them are produced by different sources, both astrophysical and non-astrophysical. For example, SN represent the waveform generated by axi-symmetric core collapse of supernovae while SG represent a Sine-Gaussian non-astrophysical glitch [9]

that facilitates automated detection of the transients from LIGO data pipeline. Of the various tools they tried, it was found that the DBNN was able to do the most reliable discrimination of all the patterns close to 100% correctness on simulated test data.

Simulated data is always tricky for the reliable estimation of machine learning models. This is because simulation itself is done using a model that has a few parameters and the machine learning algorithm learns these parameters to give apparently high accuracy. In contrast, the real signals are just representations of a much more generic phenomena and thus may have several other features that are not adequately represented by the simulation. This makes it important to test the reliability of the model on real data to know how well they perform.

To understand how well the trained machine learning algorithm could represent real-world situations, LIGO injects transients of expected SNR into the real signal and allows algorithms to search and detect them. The authors of the paper experimented their trained DBNN classifier on 1634 transient injections that had an SNR greater than 10. The classifier showed strong correlations at actual signal injection slots indicating the reliability of the model.

The interferometer is so sensitive that even the magnetic field fluctuations caused by thunderstorm and lightning will produce transient signals in the detector. The

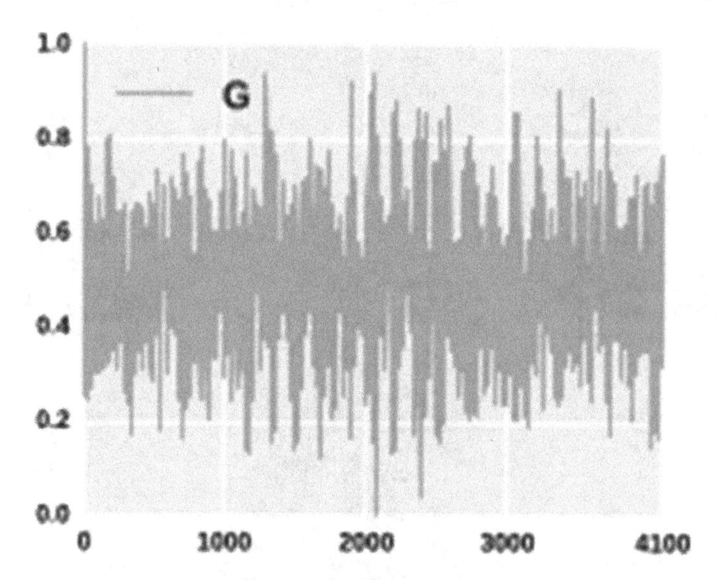

Fig. 8 Though transients are well defined, the inherent noise in the detector will be much larger than the signal making it extremely difficult for their detection

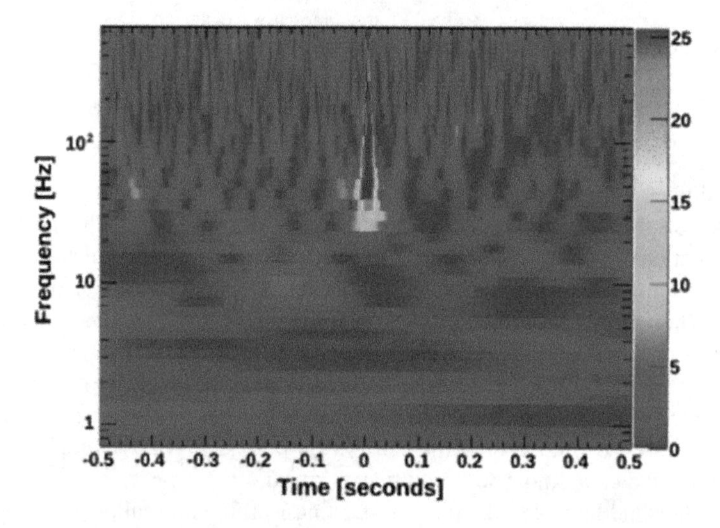

Fig. 9 Though transients are well defined, the inherent noise in the detector will be much larger than the signal making it extremely difficult for their detection

local weather stations accurately measure such events to mark those transients from being misinterpreted as gravitational wave signals. A 30 minutes real data from 16 December 2015 was used to train the classifier to isolate lightning transients from other types of transient events in the data. Two analyses were done on about 40 lighting events from the X- and Y-arms of the LIGO Livingston Observatories and

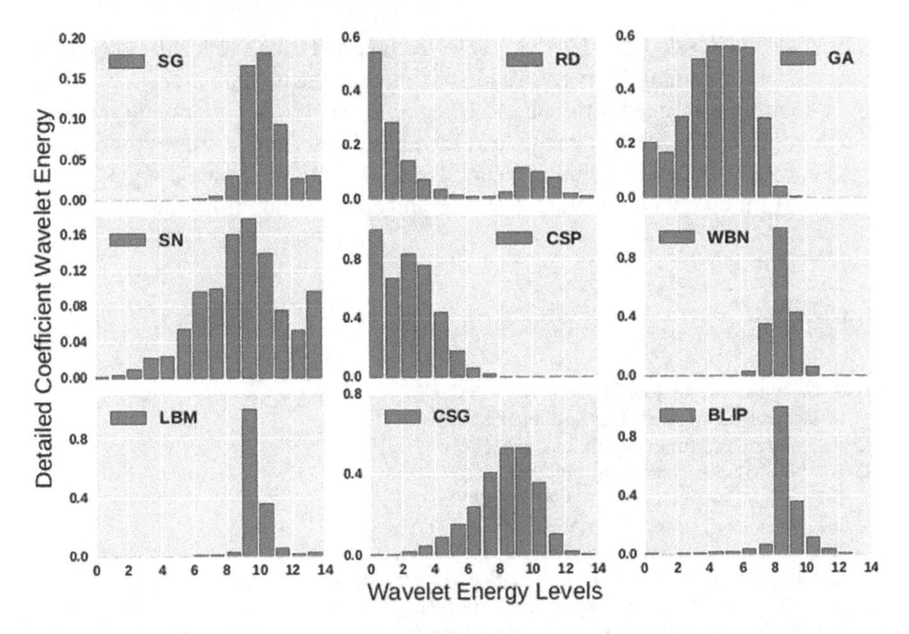

Fig. 10 Though transients are well defined, the inherent noise in the detector will be much larger than the signal making it extremely difficult for their detection

they showed that the classifier could correctly identify all except one in the Y-arm side and all except 6 in the X-arm side of the interferometer. This was the first demonstration of the use of machine learning models for real LIGO signal detection.

2 Conclusion

Machine learning is a fast-growing branch of computational logic with lot of applications in Physical Sciences. Though usually misunderstood as a branch of computer science, the author wants to emphasise that the development of two algorithms about 20 years ago in the Physics department of Cochin University under the supervision of Prof. K. Babu Joseph has applications even today in challenging problems such as gravitational wave detection and weather forecasting. Physical science is built on logic, reasoning, experimentation, prediction and evaluation. It is the same procedure that is required to build machine learning algorithms. The major difference between machine learning models from physical sciences is the absence of axioms. In machine learning, everything is assumed to be encapsulated in the data (observation) and a model is accepted as genuine if it is able to predict accurately on real data.

In all the models described in this article, features had to be extracted and supplied to the model for training. This gives a lot of control on the training process by allowing to filter out unwanted information. However, it also makes it difficult to model complex systems. Modern machine learning tools that the author currently work on uses methods such as Convolution Neural Networks(CNN), semantic segmentation and LSTM for automating feature detection and classification from row images.

References

1. S.A. Gelman, Annu. Rev. Psychol. **60**, 115 (2009)
2. S. Linnainmaa, Master's thesis, Linnainmaa (1970)
3. B.J. Wythoff, Chemom. Intell. Lab. Sys.**18**(2), 115 (1993)
4. B.J. Ninan, Neurocomputing **47**(1), 21 (2002)
5. B.J. Ninan, Comput. Geosci. **29**(2), 215 (2003)
6. B.J. Ninan, et al., Intell. Data Anal. **4**, 463 (2000)
7. K. Ninan, Astrono. Astrophys. **385**(3), 1119 (2002)
8. N. Sheelu, et al., Mon. Not. R. Astron. Soc.**419**, 80 (2012)
9. S. Nikhil, et al., Phys. Rev. D.**95**, 104059 (2017)
10. http://rasbt.github.io/mlxtend/userguide/generalconcepts/gradient-optimization
11. Simeon Kostadinov, Mirror, 8 Aug 2019
12. https://patrickhoo.wixsite.com/diveindatascience/singlepost/2019/06/13/Activation-functions-and-when-to-use-them
13. https://www.ligo.caltech.edu/image/ligo20150731e

Dynamics of Nonlinear Systems: Integrable and Chaotic Solutions

K. S. Sreelatha

Abstract Nonlinear dynamics deals with dynamical system whose behaviour is not linear as time evolves. Most of the natural systems are inherently nonlinear. To reduce the complexity of nature, all systems are approximated as linear. Dynamical systems can be modelled using differential equations. Nonlinear systems are usually classified as chaotic and integrable systems based on their easiness to find solutions of these dynamical equations. Integrable systems are exactly solvable and they possess infinite number of constants of motion. Solution for nonlinear integrable equations can be easily found. For chaotic systems, it is not possible to find exact solution. Approximate behaviour can be studied using numerical methods. But they are highly dependent on initial conditions. A very small change in initial condition can change the entire behaviour of the system. We have investigated the dynamics of integrable systems during my doctoral studies. Recently, we are investigating the use of chaos theory in cipher texting in cryptography and in biological systems. Also trying the possibility of applying chaos in economics. This article is a brief review of the work done during and after my doctoral studies.

Keywords Integrability · Solitons · 2D lattice · Nonlinear waveguides · Chaos

1 Introduction

Dynamical systems are usually modelled using differential equations. Physical systems are in general classified into linear and nonlinear systems based on their overall behaviour as time goes. Systems for which any change in its parameter causes a corresponding change in its behaviour is called a linear system. In nonlinear systems, a small change in one parameter will lead to an unexpected variation in the behaviour of the system.

Most of the natural systems are nonlinear. Any dynamical system can be defined using a collection of configurational coordinates and equations of motion obeyed

K. S. Sreelatha (✉)
Department of Physics, Government College Kottayam, Kerala, India
e-mail: drsreelathaks@gmail.com

K. S. Sreelatha and V. Jacob (eds.), *Modern Perspectives in Theoretical Physics*,
https://doi.org/10.1007/978-981-15-9313-0_12

by them. If the equations of motion of a system are known, one would like to solve them so that the dynamical variables at any time can be given as a function of initial variables and time. A dynamical system modelled using nonlinear equations of motion becomes dependent on the configuration. This type of dynamical systems are encountered in a large number of disciplines such as physical sciences, chemical sciences, engineering sciences, biological sciences and medical sciences. Based on the behaviour of solutions, nonlinear dynamical systems are classified into two: integrable systems and nonintegrable systems or chaotic systems. Systems whose dynamical equations which can be solved exactly using some methods are called integrable systems. Those for which no exact solution can be found are called chaotic systems. Chaotic systems can be studied only using numerical methods. The present article gives a review of my doctoral research work in integrable systems and some applications of optical solitons—mainly the modelling of nonlinear waveguides for the propagation of optical solitons. A small description of chaos theory and its importance is also presented.

2 Integrable Systems

One of the most fundamental, important and fascinating problems in the field of nonlinear dynamical system is to develop a general criterion which decides the integrability. The idea of integrable system becomes important after the description given by nonlinear evolution equations whose solutions represent the propagation of waves with a permanent profile known as solitons. Nonlinear dynamical systems are generally nonintegrable. But some systems of practical interest are integrable. Nonlinear integrable systems were discovered from eighteenth century onwards, but no one at that time has the real understanding of their characteristics and solutions. The first step towards the explanation of the relationship between the analytical structure of a system and its integrability was given by the Russian mathematician Kovalevskaya [1]. Her work focussed on the study of the motion of a rigid body with a fixed point from an analysis of the singularities of the solutions. Then, Korteveg and de Vries (KdV) made an important contribution to water wave problems and discovered a nonlinear model equation for the unidirectional propagation of long surface waves through uniform rectangular channel and the equation is now known as KdV equation. Now there exist several integrable partial differential equations (pdes) which can be derived using physically meaningful asymptotic techniques from a very large class of pdes.

Every integrable system has a number of special properties that hold only for integrable equations. The concept of solitons and inverse scattering technique (IST) method to find exact solutions for some nonlinear partial differential equations (pde) had far reaching influence and applications in various branches of mathematics, physics and engineering [2]. It has been established that many nonlinear wave equations have solutions of the soliton type and the theory of solitons has found applications in many areas of science. Among these, well-known equations are Korteweg the de-Vries (KdV), modified KdV, nonlinear Schrödinger(NLS), and sine Gordon (SG) nonlinear dynamical systems. These are completely integrable nonlinear dynamical equations.

The most remarkable property of integrable systems is the existence of special type of solutions called solitons. They ae localized waves that travel without much change in shape. The word *soliton* refers to solitary travelling waves which preserve their identities even after a collision. Solitons can be found everywhere; in the sky as density waves in spiral galaxies, as red spots in the atmosphere of Jupiter, in the ocean as waves bombarding oilwells and also in smaller natural and laboratory systems such as plasmas, molecular systems, laser pulses propagating in solids, superfluid He, superconducting Josephson junction magnetic system, structural phase transitions, polymers, fluid flows, elementary particles and in liquid crystals. Apart from the ubiquitous existence, the importance of solitary waves lies in their interesting properties as nonlinear waves.

Nonlinear evolution equations (NLEE) possessing soliton solutions shows many special properties such as an infinite sequence of conservation laws, Lie-Backlünd symmetries, multisoliton solutions, Backlünd transformations and reduction to ordinary differential equations of Painleve-type. Furthermore, these equations may be obtained by considering the compatibility of two associated linear operator that can be expressed in the Lax's form. All these suggest that the equations are exactly solvable. Soliton bearing equations such as sine-Gordon, KdV or NLSE are familiar to mathematicians because of the complete integrability of their Hamiltonian systems from which they derive. But when realistic applications in various fields such as condensed matter physics and engineering are considered, we have to include various perturbations which leads to problems beyond those of pure integrable systems. This was the motivation of the my Ph.D. thesis work. We studied the integrability of some perturbed nonlinear partial differential equations using Lax method and Painleve analysis.

2.1 Perturbed Nonlinear Schrödinger Equation

Among the important class of nonlinear integrable systems, the nonlinear Schrödinger equation (NLSE) plays a significant role due to the presence of the special type of stable solitary wave solutions, called envelope solitons. In optics, these solitons are expected to be suitable information carriers in optical fiber communication systems and are known as optical solitons. Solitons themselves can form a nonlinear superposition without exchange of energies; i.e. the interaction of solitons in any integrable system is elastic. This is because the associated equations possess an infinite number of conserved quantities. The simplest form of NLSE is

$$\beta U_{xx} + \gamma U|U|^2 = iU_t \tag{1}$$

where β and γ are constants. Here $|U|^2$ represents the potential which traps the wave energy that may tend to spread due to dispersion. At some values of the pulse width, the spreading effect due to nonlinearity balances and a stationary pulse or soliton can be formed. The propagation of solutions of NLSE through fiber is an important area of research for the last few years, which has experimentally proved to be an efficient way of pulse compression. Eventhough the theoretical model assumed lossless fibers, it is difficult to apply soliton propagation in practical long distance systems because the real fibers have finite losses. The first experimental observation of solitons in optical fibers was made in 1980. Now, of the research laboratories around the world, solitons are proving the key to repeaterless transoceanic optical fiber cables. Motivated by these observations, we investigated the influence of nonlinearity on the process of envelope soliton propagation. The perturbed NLSE we studied was in the form

$$ir_t + r_{xx} + 2r|r|^2 = iF \tag{2}$$

where $r = r(x, t)$ is the complex field envelope and F is small perturbing influence. This equation is assumed to represent a small perturbing influence on the propagating soliton through a non-ideal anomalously dispersive single mode optical fiber. Depending on the nature of perturbation we apply, the solution can exhibit quite complicated features. We considered a general perturbation of the form $F = i\epsilon f(x, t)r|r|^2$ where f is a real-valued function and ϵ is a numerical parameter. Then Eq. (2) becomes

$$ir_t + r_{xx} + 2r|r|^2 = -\epsilon f(x, t)r|r|^2 \tag{3}$$

We carried out the integrability studies of Eq. (3) using Painleve and Lax methods. Painleve analysis is considered to be the most powerful method for identifying integrable systems. We used the Painleve test for partial differential equations (pde) in which there is no need to reduce the pde to an ode. A partial differential equation (pde) is said to possess the Painleve property if the solutions of the pde are single-valued in the neighbourhood of a non-characteristic movable singularity manifold. With the perturbing term, the NLSE is found to pass the Painleve test irrespective of whether f depends on x and t or only on t. Hence the perturbed NLSE is found to be integrable in the Painleve sense. We obtain a Backlünd transformation also for this equation. Using Lax's method, we found that this perturbed equation possesses Lax pairs only when f depends on both x and t and subject to a certain condition. When f is time dependent only, Lax integrability fails. We also investigated the nature of the solution of this equation when the function f depends only on time. The solution of this equation is obtained using direct integration method and is given below.

$$r(x, t) = \sqrt{\frac{2k}{(1 + \epsilon k)}} \exp(ikt) \operatorname{sech}\sqrt{2k}(x - x_0) \tag{4}$$

The amplitude of the wave given in Eq. (4) depends on the strength of the perturbation, i e as the value of $f(t)$ increases, the amplitude decreases, which is as expected. Thus, we conclude from this study that existence of soliton solution in a nonlinear dynamical system is sensitive to perturbations [3].

3 Propagation of Solitons Through 2D Lattice

Research on propagation of solitons through discrete lattices dates back to the early days of soliton theory. The relevant nonlinear equations which model these lattices are very difficult to solve analytically. Generally, one looks for possible pulse soliton solutions in the continuum or longwavelength approximation. It has been found that by considering the weak nonlinear case, it is possible to reduce a large number of one-dimensional nonlinear systems to integrable ones. This nonlinear approximation has two assumptions:

(1) the amplitude of the wave is small but finite, and
(2) the wave is a long wave or a modulation of a monochromatic wave.

From recent literatures, it is found that the reductive perturbation method (RPM) is very useful for carrying out weak nonlinear approximation since it takes into account a competition between nonlinearity and dispersion in a systematic manner. The perturbation method has been developed and formulated in a general way by Taniuti and his collaborators. This method was first established for the reduction of a fairly general nonlinear system to a single tractable nonlinear equation. By using RPM, it is easy to reduce nonlinear systems to soliton equations. In the present work, we studied the wave propagation through a 2D lattice for three specific cases : quadratic nonlinearity, cubic nonlinearity and both of these together [4]. In each case, these equations reduce to three different nonlinear equations. We studied the integrability of these equations using Painleve method and also compared the solutions of these equations.

3.1 Deduction of the Nonlinear Equations for Wave Propagation Through 2D Lattice

We considered a nonlinear lattice with nonuniform mass distribution. The force between two adjacent particles is given by

$$F = k(\Delta + \alpha\Delta^2 + \beta\Delta^3 + \cdots) \tag{5}$$

where Δ is the elongation of the spring and k is the spring constant. For a particle of mass m_i and displacement a_i, the equation of motion can be given as

$$m_i\ddot{a} = K(a_{i+1} - a_i + \alpha(a_{i+1} - a_i^2 + \beta(a_{i+1} - a_i^3 + \cdots) - K(a_i - a_{i-1} + \alpha(a_i - a_{i-1})^2 + \cdots) \quad (6)$$

Assuming that the inhomogeneity is very small and does not depend on time, and making the following transformations

$$m_i = \tilde{m}(1+\rho) \quad (7)$$

$$\rho = \epsilon\rho_1 + \epsilon^2\rho_2 + \cdots \quad (8)$$

with $\tilde{m}$ is the average mass and $\rho_1, \rho_2, \ldots$ are functions of the lattice site i. Considering the lattice spacings as h in the x-direction and k in the y-direction. Then, $a_i = a_i x, y, t$ We considered the following three wave motions:

(a) Slowly varying in x, y and t for quadratic nonlinearity, $\alpha \neq 0$, $\beta = 0$
(b) Slowly varying in x, y and t for cubic nonlinearity,$\alpha = 0$, $\beta > 0$
(c) Slowly varying in x, y and t for quadratic nonlinearity along with cubic nonlinearity, $\alpha > 0$, $\beta > 0$.

When the wavelength is very large compared to the spacing of particles in a lattice, one can make the Taylor expansion on a_{i+1}. To study the wave propagation through a 2D lattice, we consider the Taylor series expansion for two variables to expand near a_{i+1}:

$$a_{i+1} = a_i + ha_x + ka_y + \frac{1}{2}[h^2a_{xx} + 2kha_xa_y + k^2a_{yy}] + \cdots \quad (9)$$

where a_x and a_y are corresponding derivatives of a_i

Case(a): Quadratic nonlinearity $\alpha \neq 0, \beta = 0$.
For $\beta = 0$, Eq. (6) becomes

$$m_i\ddot{a} = K(a_{i+1} - a_i + \alpha(a_{i+1} - a_i)^2 + \cdots - (a_i - a_{i-1})\alpha(a_i - a_{i-1})^2 - \cdots \quad (10)$$

From Eqs. (6), (7), (8), (9) and (10), we get

$$(1+\rho)\ddot{a}_i = \frac{K}{\tilde{m}}\Big[h^2a_{xx} + k^2a_{yy} + 2hka_{xy} + 2\alpha h^3a_xa_{xx} + 2\alpha hk^2a_xa_{yy} + 2kh^2a_{xx}a_y + 2\alpha k^3a_{yy}a_y + \frac{h^3}{12}a_{xxx} + \cdots\Big] \quad (11)$$

Introducing change of variables x, y and t to ζ, η and τ using the following transformations:

$$\eta = \frac{\epsilon}{h}(x - vt),\ \zeta = \frac{\epsilon^2}{k}y,\ \tau = \frac{\epsilon^3}{24h}t \quad (12)$$

where v is the velocity of sound given by $v = h\sqrt{\frac{K}{\tilde{m}}}$ Also

$$a(x, y, t) = \frac{-\epsilon}{4\alpha}\phi(\eta, \zeta, t) \tag{13}$$

Substituting equations (12) and (13), in Eq. (11), we get

$$(1 + \epsilon\rho_1 + \epsilon^2\rho_2 + \epsilon^3\rho_3 + \cdots)\left(\frac{-\epsilon^3 v^2}{\alpha h^2}\phi_{\eta\eta} + \frac{\epsilon^5 v}{48\alpha h^2}\phi_{\eta\tau} - \cdots\right) = \frac{K}{\tilde{m}}\left(\frac{-\epsilon^3}{4\alpha}\phi_{\eta\eta} - \frac{\epsilon^5}{\alpha}\phi_{\zeta\zeta} - \cdots\right) \tag{14}$$

Equating equal powers of ϵ on either sides of Eq. (14)

$$\epsilon^3 : \frac{K}{\tilde{m}} = \frac{v^2}{h^2} \tag{15}$$

$$\epsilon^4 : \rho_1\phi_{\eta\eta} = 2\dot{\phi}_{\eta\zeta} \tag{16}$$

$$\epsilon^5 : -\rho_2\phi_{\eta\eta} + \frac{1}{12v}\phi_{\eta\tau} = -\phi_{\tau\zeta} + \frac{1}{2}\phi_\eta\phi_{\eta\eta} - \frac{1}{12}\phi_{\eta\eta\eta\eta} \tag{17}$$

Using another change of variable

$$X = \eta + 12\int \rho_2(\tau)d\tau, T = \tau, Y = y \tag{18}$$

and

$$U(X, Y, T) = \phi_\eta(\eta, \zeta, \tau) \tag{19}$$

Then Eq. (17) reduces to

$$\frac{\partial}{\partial X}(U_T - 6U_{XX} + U_{XXX}) = -12U_{YY} \tag{20}$$

which can be written as

$$U_{TX} - 6U_x^2 - 6UU_{XX} + U_{XXXX} + 12U_{YY} = 0 \tag{21}$$

Thus, for quadratic nonlinearity, the equation of motion for wave propagation through 2D lattice reduces to two-dimensional form of KdV equation which is now known as KP equation.

Case(a): Quadratic nonlinearity $\alpha > 0, \beta = 0$.
In this case, the equation of motion becomes

$$m_i\ddot{a} = K(a_{i+1} - a_i + \alpha(a_{i+1} - a_i)^2 + \beta(a_{i+1} - a_i)^3 + \cdots - (a_i - a_{i-1}) - \alpha(a_i - a_{i-1})^2 - \beta(a_i - a_{i-1})^3 - \cdots \tag{22}$$

Following the same procedure as before, and equating powers of ϵ on both sides, for ϵ^4 we get

$$\rho_2\phi_{\eta\eta} - \frac{\phi_{\eta\tau}}{12v} = \phi_{\zeta\zeta} + \frac{1}{12}\phi_{\eta\eta\eta} + \frac{1}{12}\phi_\eta^2\phi_{\eta\eta} \tag{23}$$

Introducing the change of variables similar to the previous case, Eq. (22) reduces to

$$U_{TX} + U_{XXXX} + 6U^2U_{XX} + 12UU_x^2 - 12U_{XX} = 0 \tag{24}$$

We called this equations as Kadomtsev Petviashvili (KP) equation.

Case(b): Cubic nonlinearity $\alpha = 0, \beta > 0$.
Now the equation of motion becomes

$$m_i\ddot{a} = K(a_{i+1} - a_i + \beta(a_{i+1} - a_i)^3 + \cdots - (a_i - a_{i-1})\beta(a_i - a_{i-1})^3 - \cdots \tag{25}$$

Following the same procedure as before with $a(x, y, t) = \frac{1}{6\beta}\phi(\zeta, \eta, \tau)$, then equating powers of ϵ on both sides, for ϵ^4 we get

$$\rho_2\phi_{\eta\eta} - \frac{\phi_{\eta\tau}}{12v} = \phi_{\zeta\zeta} + \frac{1}{12}\phi_{\eta\eta\eta} + \frac{1}{12}\phi_\eta^2\phi_{\eta\eta} \tag{26}$$

Introducing the change of variables similar to the previous case, Eq. (25) reduces to

$$U_{TX} + U_{XXXX} + 6U^2U_{XX} + 12UU_x^2 - 12U_{XX} = 0 \tag{27}$$

We called this equation as modified KP (mKP) equation.

Case(c): Quadratic nonlinearity along with Cubic nonlinearity $\alpha > 0, \beta > 0$.
In this case, the equation of motion becomes

$$\begin{aligned} m_i\ddot{a} = K(a_{i+1} - a_i + \alpha(a_{i+1} - a_i)^2 + \beta(a_{i+1} - a_i)^3 + \cdots - (a_i - a_{i-1}) \\ * - \alpha(a_i - a_{i-1})^2 - \beta(a_i - a_{i-1})^3 - \cdots \end{aligned} \tag{28}$$

Following the same procedure with $a(x, y, t) = A\phi(\zeta, \eta, \tau)$ and equating powers of ϵ on both sides, for ϵ^4 we get

$$\rho_2\phi_{\eta\eta} - \frac{\phi_{\eta\tau}}{12v} = \phi_{\zeta\zeta} + \frac{1}{12}\phi_{\eta\eta\eta\eta} + \frac{1}{12}\phi_\eta^2\phi_{\eta\eta} + 2\phi_\zeta\phi_{\eta\eta} \tag{29}$$

Introducing the change of variables similar to the previous case, Eq. (28) reduces to

$$\frac{\partial}{\partial X}(U_T - U_{XXX} - 12U^2U_X) - 12U_{YY} = 24U_XU_Y + 24\frac{\partial}{\partial Y}\int U dX \tag{30}$$

This equation is an integro-differential equation which cannot be solved exactly.This means that as we apply the quadratic and cubic nonlinearity together, the system become more perturbes and may lead to chaotic situation. This work considered the nonlinear wave propagation through a two-dimensional lattice having nonuniform mass distribution. We used weak nonlinear approximation for quadratic nonlinearity and cubic nonlinearity seperately and both together. Using RPM, we reduced the equations of motion for three nonlinear equations, namely, Kadomtsev Petviashvili (KP) equation, modified KP equation and an integro-differential equation [4]. The integrability studies of these equations were also carried out and the details are given in next session.

3.2 Integrability Studies of KP and mKP Equations

The integrability of the derived KP, mKP and integro-differential equations is studied using both Painleve method (P-test) and Lax method. For P-test, the presence or absence of movable, non-characteristic, critical singular manifolds should be determined. When the system is free from movable critical manifolds, the P-property holds suggesting P-integrability. Main steps involved in the P-test of pdes include (i) determination of leading order behaviours, (ii) identification of powers at which arbitrary functions can enter into the Laurent series called resonances and (iii) verifying that at the resonance values, sufficient number of arbitrary functions exist without the introduction of movable critical manifolds. With quadratic nonlinearity only, the KP equation holds all these properties and it is found to be integrable in P-sense. For mKP equation with cubic nonlinearity, the resonant values do not provide sufficient number of arbitrary functions and hence it is not integrable in the Painleve sense. When both the nonlinearities were considered together, due to the presence of the integral term, it is not possible to apply the P-test and hence it does not belong to the integrable class. Hence, it has been proved that a perturbed nonlinear equation is integrable when the perturbation is homogeneous and when the perturbation is inhomogeneous, the equation becomes nonintegrable.

We also tried to find the lax pair for all these equations. Existence of Lax pair for KP equation is already proved in literature and it is considered as a completely integrable system. We could not find any Lax pair for mKP equation and also for the integro-differential equation. The non-existence of Lax pairs clearly indicates the nonintegrability of the corresponding system (in the Lax sense).

Kadomtsev and Petviashvili (KP) equation is the generalized KdV equation for the two-dimensional case. These equations model slowly varying waves in dispersive media. The equation with +ve sign ($\alpha^2 = -1$) arises in the study of plasmas and also in the modulation of long weakly nonlinear water waves which propagate in one dimension. The equation with +ve sign ($\alpha^2 = +1$) arises in acoustics and lattice dynamics. Researches into physical, earth and life sciences have led to the discovery of hundreds more nonlinear evolution equations. Of these, only a few are known to have soliton solutions.

3.3 Solitary Wave Solutions

We used travelling wave method to find the solutions for KP and mKP equations.
For KP equation, we assumed the solution as

$$U(X,Y,T) = f(\psi) \tag{31}$$

where $\psi = X + Y - cT$ Substituting this in Eq. (29) and integrating twice with respect to ψ, we get

$$-cf - 3f^2 + f_{\psi\psi} + 12f + Af + B = 0 \tag{32}$$

multiplying with f_ψ and integrating again,

$$f_\psi^2 = 2\left(\frac{f^3}{3} + (\frac{c}{2} - 6 - \frac{A}{2})f^2 - Bf - C\right) \tag{33}$$

$$\psi - \psi_0 = \int \frac{df}{\sqrt{2(\frac{f^3}{3} + (\frac{c}{2} - 6 - \frac{A}{2})f^2) - Bf - C}} \tag{34}$$

Applying boundary conditions, all the constants of integration reduces to zero and Eq. (33) becomes

$$\psi - \psi_0 = \int \frac{df}{f\sqrt{2f + (c-12)}} \tag{35}$$

which leads to the solution

$$f(\psi) = \frac{-1}{2}(c-12)sech^2(\frac{1}{2}\sqrt{(c-12)}(\psi - \psi_0)) \tag{36}$$

which on transformation gives

$$U(X,Y,T) = \frac{-1}{2}(c-12)sech^2(\frac{1}{2}\sqrt{(c-12)}(\psi - \psi_0)) \tag{37}$$

where $\psi = X + Y - cT$.

This equation represents the soliton solution for KP equation and its time evolution is given in [2, 4] (Fig. 1).

The same method and transformation equations were used to find the solution of the mKP equation. Substituting equation (30) in Eq. (26), we get

$$\frac{\partial}{\partial\psi}(-cf_\psi - 6f^2 f_\psi + f_{\psi\psi\psi}) + 12f_{\psi\psi} = 0 \tag{38}$$

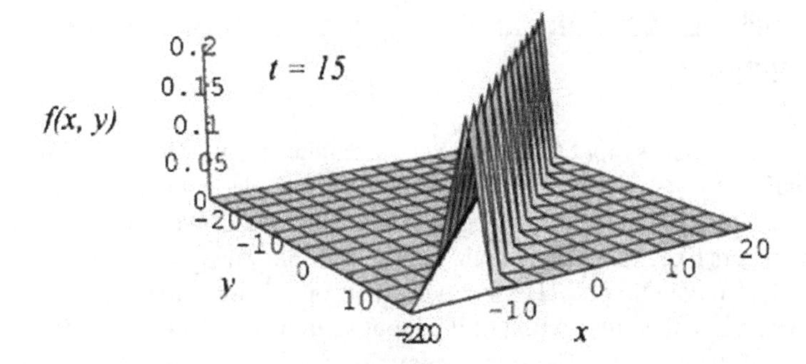

Fig. 1 Soliton solution for KP equation

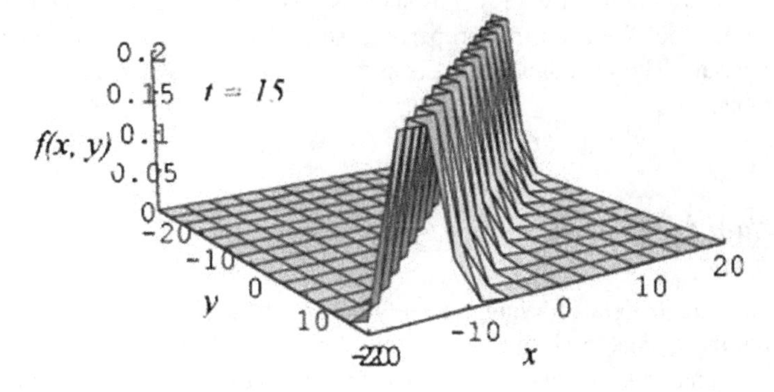

Fig. 2 Soliton solution for mKP equation

Using the same integration techniques as in the previous case, we reach at the solution

$$f(\psi) = -(c-12)sech(\sqrt{(c-12)}(\psi - \psi_0)) \quad (39)$$

This gives the solitary wave solution for mKP equations and its graphical representation is given in Fig. 2.
This section discussed the propagation of wave through a two-dimensional lattice. From the equation of motion, we derived the modelling equation for wave propagation for three cases: quadratic nonlinearity, cubic nonlinearity and their combination. The first two cases we found soliton solutions and for the third case we could not find a soliton. Thus, we conclude that the lattice dynamic system with quadratic nonlinearity is completely integrable while the system with cubic nonlinearity is partially integrable. When the quadratic and cubic nonlinearities act together, the system is not integrable and may become chaotic. As expected, nonlinearity in evolution equations may cause different behaviours, from regular motion to chaotic motion.

4 Modelling of Nonlinear Waveguide for Optical Soliton Propagation

The use of solitons as input for nonlinear waveguides is investigated during the doctoral studies of my student. We have modelled waveguides using nonlinear materials such as ZnO, MgO and TiO_2 with air as upper cladding. We have optimized various parameters like input wavelength, thickness of the waveguide, core nature of the waveguiding medium and pulse width of the propagating soliton. For a dispersion free propagation, the pulse width of the input soliton should be compatible with the thickness of the core [6]. The wide bang gap of the core materials can be utilized for wavelength and band gap tuning. The refractive index of these materials can be enhanced by doping them with metals such as gold which will enhance their optical properties too. We used this property to study the propogating modes of high index materials. These studies will be beneficial to the fields of MEMS and sensing applications.

4.1 Planar Waveguides

Nonlinear optic materials for integrated nonlinear optics propose stringent problems as regards their process ability, adaptability and interfacing with other materials. These additional requirements are intrinsically related to the fabrication of nonlinear devices, which besides efficiently performing the expected nonlinear operation, must be miniaturized, compact and should be reliable. The magnitude and speed of the nonlinearities are essential characteristics in any assessment of the material for NLO devices.

In recent years, nonlinear optics has grown all over the world into one of the most important research area. The significance of metallic oxides in the fabrication and the creation of new materials for industry have resulted in a tremendous increase of innovative waveguide processing technologies. Currently, this development goes hand-in-hand with the explosion of scientific and technological breakthroughs in devices like photonic crystal fibers, and as the root key to the phenomena of Supercontinuum. The tunability as mentioned earlier can put a new dimension to the concept of such white lasers. Further research is on the thickness of the thin film waveguide structures which promise to yield innovative results. We concentrated on the experimental route to fabrication of oxide waveguides for nonlinear applications. The modelling is done using software Matlab [5, 6].

A thin film planar waveguide consists of three layers of different dielectric materials with the central guiding layer having a refractive index greater than both outer layers. The upper layer is usually air, an additional layer of dielectric material may be deposited on top of the thin film, which act as the cladding. The principle of light confinement is total internal reflection, exactly the same as in an optical fiber. If a guiding layer with refractive index n_0 and width $2L$ is sandwiched between two (thick) layers of refractive index n_1, the number of modes the waveguide can support depends on the waveguide parameter

$$V = 2\pi \frac{2L}{\lambda} \sqrt{n_0^2 - n_1^2} \tag{40}$$

where λ is the wavelength of the light. For total internal reflection, we require a high index material surrounded on the top and bottom by lower index materials. The requirement for a wave to propagate through the waveguide is that the angle of incidence must be greater than the critical angle, $sin\theta_c = \frac{n_2}{n_1}$ where n_1 and n_2 are the refractive index of the core and cladding, respectively.

For an asymmetric waveguide the cut-off frequency above which the multiple modes of a particular polarization can propagate is given by

$$\omega_c = \frac{\pi_c}{d} \frac{1}{\sqrt{n_1^2 - n_2^2}} \tag{41}$$

The number of modes in a wave guide should always be minimal for the ease of designing and fabrication. We consider a waveguide single mode which supports a single-guided mode for a particular polarization (TE or TM).

4.2 Optical Solitons and NLSE

The pulse propagation within a nonlinear waveguide needs a spatial restriction to reduce the effective area and enhance nonlinear interactions. This helps to reduces the dispersion thereby controlling the spreading of the pulse. This introduces an interplay between the nonlinearity and the group velocity dispersion which leads to the formation of optical solitons [5, 6]. The dispersion length and nonlinear length where such effects may become dominant are given by $Ł_0 = \frac{L_0^2}{|\beta_2|}$ and $Ł_N L = \frac{1}{\gamma P_0}$, respectively. When $L_D \leq L$ and $L_{NL} \geq L$ the pulse undergoes significant dispersive broadening, but the spectrum remains constant. If $L_D >> L$ and $L_{NL} \leq L$, the spectrum will change via self-phase modulation due to nonlinear effects. When both $L_D \leq L$ and $L_{NL} \leq L$ the interplay between the dispersion and the nonlinearity produces different results depending on relative signs of the two effects. If the dispersion is anomalous with $\beta < 0$ and $n_2 > 0$, stable solutions, known as solitons, forms. Soliton solutions are extremely stable because they can shed excess energy in the form of a dispersive wave until a stable solution is formed. Inorder to propogate solitons, the waveguide parameters are to be fixed [6]. These parameters are fixed using numerical methods. By solving the wave equation and computing the modal eigenvalue of a waveguide structure, the modal index or the effective refractive index may be calculated as

$$n_{eff} = \frac{\beta}{k_0} \tag{42}$$

$$k_0 = \frac{2\pi}{\lambda} \tag{43}$$

The normalized refractive index is calculated as

$$b = \frac{n_{eff}^2 - n_2^2}{n_1^2 - n_2^2} \tag{44}$$

An ideal propogation requires

$$n_2 \leq n_{eff} \leq n_1, 0 \leq b < 1 \tag{45}$$

provided $n_2 > n_1$.

We considered ZnO waveguides for soliton propagation. Numerical analyses of the ZnO waveguide structures were carried out. The field distributions were plotted and given in Fig. 3 using the software Mathworks wavemode solver. The dependence of the plotted field distributions with the input wavelength and the refractive index was justified, assuming that the refractive index of the waveguide structure consists of ZnO with a refractive index value, $n_2 = 1.975$. The silica substrate lower refractive index was taken to be $n_1 = 1.456$. The waveguide structure was completed by considering a lower refractive index of the air upper cladding, $n_0 = 1.000$. The field modes were plotted in the $800 - 1200$ nm range for the input wavelength where intense nonlinear effects and dispersions were observed [7].

Optical nonlinearity of metal nanoparticles in a semiconductor has also attracted much attention because of the high polarizability and fast nonlinear response that can be utilized in making them as potential optical devices. Out of various metal nanoparticles, silver, copper and gold are extensively studied in colloids, thin films and in different glass matrices for their nonlinear optical properties. In our work, we have chosen silver nanoparticle doped with ZnO, because of their interesting optical properties in the visible range which gives rise to wide applications in optoelectronic devices. The propagation of soliton pulse through a doped ZnO core of higher refractive index have been plotted in Matlab software and is reported [6]. The elliptical field distributions obtained enable a solitonic wave profile of amplitude, $A(x) = A_0 sech\frac{x}{a}$. Here the radius of the core is taken to be $0.0673\mu m$ to enable nonlinear effects.

Numerical analysis of ZnO, MgO and TiO_2 waveguide structures were also carried out. The dependence of field distributions with input wavelength and refractive index was plotted. ZnO with a refractive index value, $n_2 = 1.975$ and MgO with a lower refractive index of 1.7375 were considered initially. A considerable increase in a doped structure of ZnO with silver, $n_2 = 2.0037$ was compared with that of TiO_2, $n_2 = 2.49621$. The silica substrate with a lower refractive index was taken to be $n_1 = 1.456$. The waveguide structure was completed by considering a lower

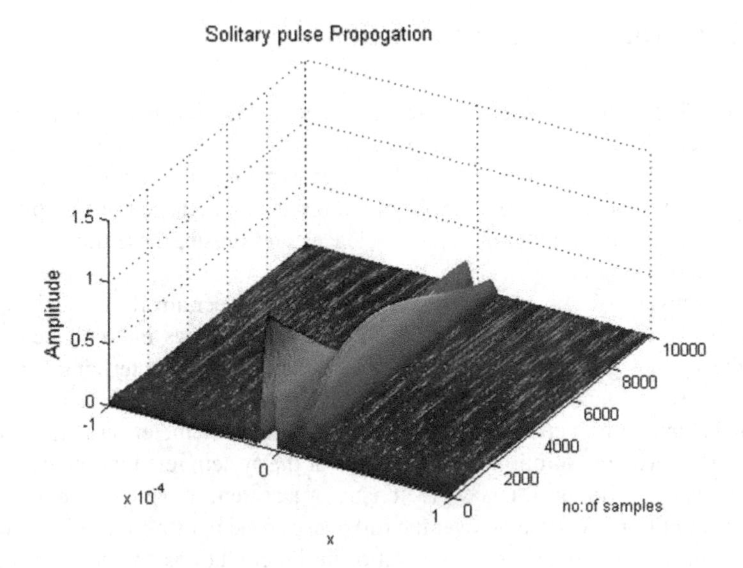

Fig. 3 Soliton propagation through ZnO waveguide

refractive index of air as the upper cladding, $n_3 = 1.000$. The field modes were plotted for $400 - 700$ nm of the input wavelength where intense nonlinear effects and dispersion were observed. In the case with TiO_2 having $n_2 = 2.49621$, it exhibited a lower dispersive regime when compared to the undoped variants. TiO_2 can thus prove to be a worthy candidate for fabrication of nonlinear waveguides which can facilitate the passage of solitary pulses [15].

In this work, we modelled doped oxide Waveguide configurations that enabled the passage of a soliton in due to the effect of the increase in refractive index with doping. The width of the waveguide may be below 6 um for soliton propagation which should be compatible with the input wavelength. Higher index cores such as TiO_2 showed better focusing for the soliton propagation when compared to lower index cores such as MgO. For triangular index guides, the oscillatory nature of the input soliton may be controlled to a certain extent by using a soliton solution with a width factor; the width may then be optimized (Fig. 3).

If solitons can be experimentally propagated through such cost-effective waveguides, it would undoubtedly be a boost to the optical communication scenario. The information can then be transferred at a quicker rate without loss. The information carrying capacity of solitons may be used for the fabrication of various communication channels and switching networks as well. The study of evanescent waves such as evanescent sensing and their applications in various multidisciplinary optical fields may also be consequently studied.

5 Chaos Theory—An Introduction

In the first session of this article, I have mentioned that nonlinear dynamical systems can in general be classified to integrable systems and chaotic systems. My research work in the area of integrable systems has been explained in the previous sections. In this section, I will briefly explain about chaos theory and its potential applications in economics, cryptography, biological sciences and music; these are the areas in which I am working now [8].

The discovery of apparently random behaviour of certain dynamical systems turned out to be quite revolutionary leading to many issues interconnecting stability theory, new geometrical features and new signatures characterizing dynamical performances. Systems which are basically nonlinear and exhibiting an apparently random behaviour for certain range of values of system parameters are referred to as chaotic. However, the solutions or trajectories of the system remain bounded within the phase space. Mixing: It is a characteristic of a system in which a small interval of initial conditions gets spread over the full phase space in its asymptotic evolution. In a chaotic system, an arbitrary interval of initial conditions spread over the part (attractor) of the phase space to which the trajectory asymptotically confines. Thus any region gets into every other region of the spatial attractor of phase space. The chaos theory, also called the complexity theory is a scientific discipline which is based on the study of nonlinear systems. Chaos theory can be considered as a mathematical method that allows us to extract beautifully ordered structures from a group of chaotic systems—complex natural systems such as the beating of the human heart and the trajectories of asteroids.

5.1 *Lorentz System*

A meteorologist named Edward Lorenz in 1961 made a profound discovery while trying to find methods to predict weather using computational techniques. It was a continuous time nonlinear system exhibiting chaotic trajectories for specific values of system parameters. The system consists of a set of three ordinary differential equations to model a thermally induced fluid convection in the atmosphere given by

$$\frac{dx}{dt} = \sigma(y - x); \frac{dy}{dt} = Rx - y - xz; \frac{dz}{dt} = xy - \beta z \tag{46}$$

These equations are well known as Lorenz equations. This model could explain the uncertainties observed in weather predictions. The phase space of this system can be plotted and the event is being referred to as *Butterfly effect*. According to Lorenz, unpredictability in complex systems is called "sensitivity to initial conditions". This means that, in a complex, nonlinear system, a tiny difference in starting position can lead to greatly varied results. In his own words "if a butterfly is flapping its wings in Argentina and we cannot take that action into account in our weather prediction, then we will fail to predict a thunderstorm over our home town two weeks from now because of this dynamic" [9]. Precise dynamical properties of this system are exploited for studying various physical, biological, chemical and environmental systems and also in cryptographic schemes. The significance of sensitivity to initial conditions is that if we start with a limited amount of information about the system then after a certain time, the system would no longer be predictable. This does not mean that one cannot assert anything about events far in the future.

To study how a system pass from being nonchaotic to being chaotic as some parameter of the system is varied continuously, there are certain parameters could be checked for the given system known as route to chaos. As an example, for any constant vector field on the three-dimensional torus, we can choose a small perturbation which results in a chaotic attractor.

Chaos can be seen in many systems such as electrical circuits, planetary bodies orbiting each other, fluid dynamics and in chemical reactions. But many real systems such as the weather need too many parameters to analyze precisely with computers, which makes these systems chaotic. Important routes to chaos are the period-doubling cascade, intermittency, crisis and quasiperiodic routes. Most commonly used model to study the dynamics of a nonlinear system is the logistic map. It uses a nonlinear difference equation which maps at discrete time steps. This model was first used to map the population value at any time step to its value at the next time step:

$$x_{n+1} = \lambda x_n(1 - x_n) \tag{47}$$

Using this model, it is easy to predict chaotic behaviour of a system by drawing a graph between λ and x_n values. Using these equations, it is possible to study the dynamics of GDP variation in different countries so as to check whether the most suitable prediction is possible in the case of GDP growth for a country [8]. The use of nonlinear oscillator equation for creating cyphertext in cryptography for more simple and secure data transfer is also studied [9]. Similarly, the phase space study of Lorenz equation can be used to model the dynamics of brain and hence we can give a prediction regarding the behaviour of an epileptic person [10–12]. We have recently started these studies and expect that we could find a good result which can be utilized for public reference.

6 Conclusion

This article mainly includes a review of the research work I have carried out in the area of nonlinear dynamics during my Ph.D. under the mentorship of Dr. K. Babu Joseph. We studied the effect of perturbation in the integrability properties of some nonlinear dynamical systems, mainly nonlinear Schrödinger equations having optical soliton solutions and Kadomtsev Petviashvili equations that model the wave propogation through 2D lattice. The results were published in reputed journal and are given in reference list [3, 4]. The research carried out by my Ph.D. student, Rosmin

Elsa Mohan, related to this area are also included. We studied the modelling of silver-doped ZnO thin film waveguide for propagation of optical solitons. The results we obtained are also published [15]. Now we are concentrating on the applications of chaos theory in economics, biology and also in music [13, 14]. Applying chaos theory in making economic predictions provide results that are closely related to traditional forecasting methods using chaos. Research work in these areas are progressing and we are waiting for good results to get published at the earliest.

References

1. M. Lakshmanan, S. Rajasekhar, *Nonlinear Dynamics* (Springer, Berlin, 2003)
2. R.K. Bullough, P.J. Caudrey, *Solitons* (Springer, Berlin, 1980)
3. K.S. Sreelatha, K. BabuJoseph, Chaos, Solitons, and Fractals, **9**(11), 1865 (1998)
4. K.S. Sreelatha, K. BabuJoseph, Chaos, Solitons & Fractals, **11**(5), 711 (2000)
5. R.E. Mohan, K.S. Sreelatha, M. Sivakumar, A. Krishnashree, Adv. Mater. Res. **403**, 3753 (2012)
6. R.E. Mohan, M. Sivakumar, K.S. Sreelatha, Mater. Res. Bull. **69**, 131 (2015)
7. K.S. Sreelatha, L. Parameswar, K.B. Joseph, AIP Conf. Proc. 1004, **1**, 294 (2008)
8. R.M. Goodwin, *Chaotic Economic Dynamics* (Oxford University Press, 1990)
9. S. Skaria, I. Jinchu, V. Jacob, K.S. Sreelatha, Int. J. Appl. Eng. Res. **15**(5), 524 (2020). ISSN 0973-4562. https://www.ripublication.com/ijaer20/ijaerv15n5_14.pdf
10. G.L. Baker, J. Gollub, *Chaotic Dynamics: An Introduction* (Cambridge university press, 1996)
11. A. Babloyantz, J.M. Salazar, Nicolis, Phys. Lett. A **111**, 3, 152 (1985)
12. Babloyantz, *Dynamics of Sensory and Cognitive Processing by the Brain*, vol. 196 (Springer, Berlin, 1988)
13. J. Harley, Chaos and music. MIT Press **28**, 221 (1995)
14. R. Jonathan, *Chaos in Music: Historical Developments and Applications to Music Theory and Composition*, The University of North Carolina at Greensboro (2009)
15. R.E. Mohan, *Numerical Modelling of II-IV Planar waveguides for Nonlinear Applications*, Ph.D Thesis (2014)